ECONOMIC COMMISSION FOR EUROPE
Geneva

Ground-water legislation in the ECE region

*A report prepared under the auspices
of the ECE Committee on Water Problems*

UNITED NATIONS
New York, 1986

The designations employed and the presentation of material in this publication do not imply the expression of any opinion whatsoever on the part of the Secretariat of the United Nations concerning the legal status of any country, territory, city or area or of its authorities, or concerning the delimitation of its frontiers or boundaries.

ECE/WATER/44

UNITED NATIONS PUBLICATION

Sales No. E.86.II.E.21

ISBN 92-1-116372-2

00950P

CONTENTS

PREFACE

Since drawing up the ECE Declaration of Policy on Prevention and Control of Water Pollution, including Transboundary Pollution, the Committee on Water Problems has devoted further attention to ground-water management. Indeed it has become a priority issue in the programme of work.

Throughout the ECE region, ground water is valued for its high quality. Availability close to the user and favourable distribution over time make it a precious resource. Ground water is therefore being used extensively and sometimes being misused by over-exploitation which has led in a number of cases to its depletion. Ground water is also endangered by certain practices leading to serious quality deterioration.

It is now realized that problems with ground water in terms of both quantity and quality are continuing to grow apace with the rising demand for good quality water. Such problems are exacerbated by low investment in ground-water exploitation compared with investments for surface water abstraction. Greater intensity of land use and agriculture, increasing industrial activities, including mining and problems of waste disposal, along with risks of accidental spills have all taken their toll. Once contaminated or depleted, ground water may be permanently impaired. This could have far-reaching and unpredictable implications for humankind. Protection strategies and remedial measures that are effective will have great significance for conserving this vital resource. The rational and wasteless use of ground water is thus being promoted as a solution for sustainable economic development.

In the light of the importance of ground water as a prime source of drinking water and mindful of the hazards of pollution and depletion with respect to maintaining this resource for present and future generations, the Committee on Water Problems convened a Seminar on Ground-water Protection Strategies and Practices, in Athens (Greece). This Seminar was one of the contributions to the implementation, at the regional level, of the International Drinking Water Supply and Sanitation Decade. The Seminar served as a forum for the exchange of views among ECE Governments on experience in the formulation, implementation and evaluation of ground-water protection strategies and related practices. Major elements of national ground-water policies and practices, such as special legislation, regulations, investment policies, socio-economic measures, institutional structures as well as policies regarding land-use planning and integrated management of surface and ground water were examined, with regard to both quality and quantity.

The Seminar was timely in that it provided an opportunity to draw attention to a resource which by its very nature is hidden from view and hence risks mismanagement. The Seminar gave momentum to efforts among ECE Governments in promoting sustainable use of ground-water resources, including their efficient protection against pollution hazards, in the context of overall water policy. The Seminar Proceedings have been published by the Greek Ministry of Energy and Natural Resources.

Regarding ground-water legislation, the Seminar concluded that "Legal provisions for the proper and sustainable management of ground water are in many countries scattered over a wide variety of laws and regulations. Relatively few countries have a legal framework designed especially for ground water. More emphasis should be given to the further development of a set of rules defining the rights and obligations of all those concerned, public authorities, industry, agriculture, environment and the general public, in using and protecting ground-water resources.".

As a follow-up to the Seminar, studies on ground-water legislation, policies, and strategies were undertaken under the auspices of the Committee on Water Problems. These studies were prepared by government rapporteurs and the secretariat. The contributions to the study on ground-water policies by Mr. P. Karakatsoulis (Greece), Mr. Lyalko (Ukrainian SSR), Mr. Margat (France) and Mr. Zwirnmann (German Democratic Republic), rapporteurs, should be adknowledged in particular. Credit is also due to Mr. Dante A. Caponera, consultant to the secretariat, who assisted in compiling the report on ground-water legislation. The financial support of UNEP to the secretariat must be mentioned with appreciation.

The Committee at its seventeenth session in 1985 endorsed for general distribution the reports on ground-water legislation and policy. It should therefore be noted that the information contained in this document reflects the situation prevailing in November 1985. In accordance with established practice, the document is published under the sole responsibility of the secretariat.

As part of its further efforts to deal with problems of ground-water management, the Committee on Water Problems will convene a Seminar on Protection of Soil and Aquifers against Non-point Source Pollution. The Seminar will take place in Spain at the invitation of the Government of Spain. Policy and legal measures to combat diffuse pollution will be the focus of this Seminar, giving special attention to land-use planning measures and the economics of soil and ground-water protection. Also the Committee is now finalizing a set of principles on ground-water management, with a view to providing guidance to ECE Governments in formulating, adopting, reviewing and implementing policies on the national and international level for the protection of ground water against pollution, over-exploitation and wastage.

INTRODUCTION

Long ago, ground-water aquifers were exploited mainly for drinking water and for agriculture. Increased water demands - in general owing to higher standards of living and advanced technology - have led to tremendous withdrawals of ground water. This is so for all economic sectors.

The reasons are technical, economic and legal. Pre-eminent among them are the availability of ground water close to users, good quality compared to other water resources, and consistency in terms of quantity and quality within one aquifer. Ground water has been a "cheap" and easy resource to exploit in the absence of legal restrictions. In addition to increased extraction, the uncontrolled discharge of waste water as well as the handling and storage of chemicals and the disposal of solid wastes pose potential pollution hazards to aquifers. The development of industrialized agriculture, with mass applications of chemical fertilizers and pesticides, has brought a new dimension to ground-water pollution. It is now widely accepted that this worsening situation demands effective control measures, supported by sound ground-water legislation and adequate monitoring.

Only recently, vis-à-vis the behaviour, characteristics and limitations of ground water - a resource which is hidden from view - has the level of understanding of this resource by policy-makers and the public at large reached a point where there is a sufficient base for propounding regulations cognizant of the exigencies of modern water management. The concept of planning, rational management and regulation of ground water on an integrated basis is now advocated by water specialists and generally accepted by legislators.

Water economics call for joint consideration of both surface and underground water resources and their sustainable development. However, special measures are necessary to protect ground water since managerial instruments, in particular legislation and regulation, cannot be directly transposed from surface to subsurface waters. Aquifers lying underneath frontiers between ECE countries pose an additional challenge for equitable and sustainable use and joint protection against over-exploitation and pollution.

One of the first difficulties for regulation of ground water relates to the very definition of ground water. Ground waters may encompass all the waters existing below the surface of the land, as reflected in the legislation of several Canadian provinces (Alberta, Manitoba, Saskatchewan). In the legislation of Finland, ground water is defined as "all the water contained in the ground base or rock base". The Water Resources Act promulgated in the United Kingdom in 1963 defined ground waters as those waters contained in "underground strata". A similar but more restrictive concept is embodied in the legislation of Turkey, which defines a "ground-water sheet", as being "every deposit of ground waters existing in the underground strata the withdrawal of which at any point exercises an effect on the total water mass". Definitions have an intrinsic importance as they serve to demarcate the resource.

From a legislative point of view, the question may arise of whether to include all subsurface water in the term "ground water" or to consider only the zones of saturation. In some legislation, an aquifer is regarded technically as a "geological formation or structure" that stores and

transmits water in sufficient quantities to supply pumping wells or springs.
Thus some water-bearing strata of negligible yield are excluded. In other
laws, an aquifer relates to all subsurface formations containing material
saturated with water, regardless of yield. Some avoid making any distinction
and identify ground waters with the aquifers containing them. Still others
attach economic concepts and distinguish between aquifers that are
potentially, technically and economically usable; they may even introduce an
ecological dimension. Further differences in national legislation relate to
whether springs, either concentrated, diffuse or subaqueous, and whether bank
filtration pertain to ground water or not.

The vast array demonstrates the need for clear definitions and common
terminology to avoid ambiguity and thus facilitate the implementation of legal
provisions in modern legislation. Many countries strive towards harmonization
of terminology and elaboration of uniform definitions, not only on a national
but also an international level. Despite these efforts, consistency in
defining ground water is lacking, especially at the international level.
Ground water was defined by the International Organization for
Standardization (ISO) in 1980 as "water which is being held in and can usually
be recovered from, or via, an underground formation" (ISO 6107/1;
term 1.1.3). In 1979 the Council of the European Communities defined ground
water as "all water which is below the surface of the ground in the saturation
zone and in direct contact with the ground or subsoil" (Council Directive on
the Protection of Ground Water against Pollution caused by Certain Dangerous
Substances).

Before drafting a national law or international agreement, it is
necessary to have a clear idea of the object to be regulated. The very nature
of ground water renders this task difficult. First, aquifers do not respect
administrative and political boundaries created by man. Often aquifers are
superimposed one upon the other, with different behaviour as to flow, quantity
and quality. In addition, aquifers may not belong to the hydrological cycle
and therefore may not be renewable. Different types of legal provisions
governing each type of ground water might be one way of coping with this
diffuse resource. Unlike surface waters, ground waters by their very nature
defy any notion of clarity. While modern technology proceeds rapidly towards
a better knowledge of ground water, legislation often lags far behind.

Implementation of legal provisions is also an important aspect. In many
countries different authorities manage and control separately ground water and
surface water. Those who consider ground water to be a mineral resource feel
that it should be governed by legislation covering minerals and administered
by institutions responsible for mining activities. Others contend that since
most ground waters are part of the hydrological cycle, they should be
regulated by general water legislation and administered by water
institutions. Such issues are still being debated in many countries.

An example of Spanish water legislation may be representative of the
stiuation in a number of ECE countries. In Spain all surface waters had been
managed since the beginning of this century at the river basin level while
ground waters, in view of their minor relevance at the time of the enactment
of the Spanish Water Law of 1879, were left out of the jurisdiction of surface
water management. While this issue is still open in some countries, there is
a move to recognize the interdependence between ground water and surface water

and to reflect this link in water legislation, this was done, for example, by the United States National Water Commission in 1973. The same concern is being considered in the new draft water law of Spain.

. Another difficulty arises when adapting and formulating ground-water legislation. It is the problem of ownership or legal status of ground-water resources. According to national legislation in a number of ECE countries, the traditional legal system had vested in the landowner the ownership of whatever was located on or below the land, including ground water. This conferred on the owner the unrestricted right to use, or abuse, the resource unlike the case of surface waters, as common belief held that no one had unrestricted rights over surface waters, even on private property. Visibility may be the key to this disparity, as surface waters may be seen flowing from one property to another. In the case of ground water, its presence is not evident to the lay person, and the population at large does not realize that one party's exploitation of ground water may directly or indirectly influence this resource in a neighbour's land.

The foregoing applies particularly to activities causing depletion of aquifers and degradation of ground water quality. Legal mores are difficult to modify, in spite of subsequent legislation introducing controls or detailed regulations. Indeed, the population at large has seldom been aware of the nature, behaviour and characteristics of ground water. This has placed a constraint on the implementation of any legislative measures tending to control activities affecting ground water.

Today's better understanding of the nature, characteristics and behaviour of underground water provides a technically sounder base on which to elaborate ground-water regulations. However, the adaptation and formulation of ground-water legislation should not be delayed while awaiting definitive scientific proof or explanations of all phenomena occurring in or around aquifers. Rather laws should be drafted on the basis of available knowledge and enforced as framework legislation flexible enough to respond expediently to conflicts and problems posed by over-exploitation and pollution of aquifers.

I. OWNERSHIP STATUS AND RIGHTS OF USE

One of the most important legal issues for integrated management of ground water and efficient protection against over-exploitation and pollution is the status of ownership and rights for rational resource use. In many countries, ground water has long been considered an individual property allowing a landowner the possibility to exercise full powers over its use. Governments are coming to realize that the water beneath the surface of the land constitutes a common resource, to be utilized by users taking into consideration the benefit of all. With this growing awareness has come a greater appreciation of the need for control and management by public authorities. Systems of acquisition of ground-water rights have been established. While they differ from one country to another, their objective is to grant use rights recognized by law or custom according to clearly established rules.

These questions and the changes of relevant concepts in various countries provide a good example of how countries with different historical backgrounds, different socio-economic systems and thus, different legal concepts, arrive at similar policies and managerial tools when confronted by the same problems: i.e. resource depletion and pollution hazards endangering ground water. This will be illustrated in more detail in the following sections.

A. Origins of ground-water legal systems

In the ECE region, existing ground-water institutions, administration and laws are derived from one or more of the following legal systems: Roman law, with its derivations; civil law and the Common Law of England; the American appropriation doctrines; doctrines applied in socialist countries; and Moslem water law. Mention should be made also of customary law which still applies at the local level in some countries.

Roman water law principles are still relevant, since, initially, they were imposed during the period when the Roman Empire flourished. They were subsequently introduced widely through the continental "civil law" systems, such as those of Belgium, France, Italy, Portugal, Spain and the Netherlands, where Roman water law constituted the fundamental base. The same holds true for ECE countries which adopted the "Common Law of England", itself influenced by Roman water law principles.

In view of its influence on subsequent water legislation, it is important to underline that Roman law divided water into three categories: (a) private, occurring within or under one's private land, subject to unlimited and unrestricted right of use by the landowner; (b) common, which did not allow any ownership of running water which could be used by anyone for any purpose without limit or permission; (c) public (large rivers, lakes and canals), owned by the State or other public institution.

B. Ground-water legislation in "civil law" or "code system" countries

According to the "civil law" or "code system", as codified by the Napoleonic Code and as found in the water laws of a number of countries, a dual system of ownership was applied: public and private. If the overlying land was in the public domain, ground water was public and subject to administrative permit for its use; if the land was privately owned, the same water and its use was private.

Ownership of the land thus conveyed ownership and use-rights over water found under the land either in conjunction with ownership rights over the water, as in Austria, Belgium, France, Greece, Portugal, Spain, or regardless of it as in Finland, Italy, the Netherlands and Turkey. As a reflection of the customary association of land and ground-water resources in law, the dichotomy public/private lingered on in the above countries, the public or private ownership status of the land generally, but not always, being the determining factor of the public or private domain status of ground water lying under it.

Under the pressure of modern technical and economic requirements and the increased concern posed by water demands and quality control over water resources, the category of "private waters" is slowly disappearing in European countries, with a greater degree of government control over private waters (surface and underground) and their uses. As an example, the recent water laws and subsequent regulations of France and Italy have, in practice, brought all water use, both public and private, under the control of either the central Government or other regional or basin authority. This trend is also appearing in other "code" countries, such as Belgium, the Netherlands, Spain, et al. Water is no longer subdivided into "public and private" but into "domaniale and non-domaniale" water - leaving aside the expression "private water". Also, the obsolete concept of "navigability and floatability" to identify public waters is being discarded.

Virtually in all the countries mentioned above, the use-rights in water laying beneath one's land have undergone a process of steady curtailment in the interest of protection of the resource against depletion. As a result, in broad outline, (a) the owner of the land now has an unrestricted right to withdraw ground water beneath that land only for domestic consumption by his or her own household; (b) all other abstractions of ground water are subject to permit and ancillary reporting requirements. This model is subject to qualifications, however, as will be seen below.

C.   Ground-water law principles in "common law" countries

English law continued to apply the old Roman concept of "common" water whereby there was no ownership of water, not even by the State or the Crown. Waters were regarded as transient and elusive for use by the riparian landowner, provided such use was reasonable. Only water (including ground water) which accumulated or fell on one's land and was collected in artificial or natural ponds and reservoirs could be privately owned; such ownership lasted only during the time of possession. The same principle applied to underground waters, which became the property of whoever abstracted and retained them in his possession. Only a court decision or an administrative ordinance or regulation could limit or restrict the right to use water by a riparian owner, over and above the limitations inherent in the "natural flow and reasonable use" criteria.

This system of water legislation has influenced the water laws and institutions of Cyprus, Malta and the United Kingdom in Europe, and Canada and the eastern States of the United States of America in North America. In these countries, underground streams were usually distinguished from all other forms of underground waters (subterranean lakes, pools, artesian water, and seeping or oozing water). Like the riparian rights doctrine the former were not susceptible to private ownership, as they were considered surface streams. Concerning the latter waters, which were lumped together under the term "percolating water", common law gave full ownership and right of use to the holder of title to the overlying land.

As in the case of the "code countries", under the pressure of modern economic and other constraints, this system has encountered strong limitations, even in the United Kingdom where the system originated. Under the Water Resources Act promulgated in the United Kingdom in 1963, a permit or licence is necessary for the abstraction and use of underground waters. The occupier of land may, however, use underground water for his or her own domestic consumption without a licence. It is worth noting the the curtailment of common law rights to ground waters was accomplished in England and Wales without actually attributing public ownership status to the resource.

In Malta and Cyprus, in view of the vital importance of ground water and the lack of permanent surface streams, all underground waters have been given public ownership status with their use subject to government permit. In the riparian rights jurisdictions of the eastern United States, a landowner's privileges have been increasingly curtailed by court-developed rules of reasonable use or replaced by a permit system.

D. American doctrines

While the original English common law asserted itself in the eastern part of the United States and in Canada as the "riparian doctrine", in some western States of the United States as a consequence of the discovery of gold, which generated the use of water for mining purposes, the "appropriation doctrine" was developed. This system accorded a water right to the first who used water according to the principle, "first come, first served - first in time, first in law".

Within the United States, this doctrine collided both with the riparian doctrine of the eastern States and with the Mexican water law principles prevailing in some western States as derived from Spanish water law and based on the "code system". Against this background, the courts in the United States have limited the original concept of the appropriation doctrine and developed new theories of "correlative rights" and of "beneficial uses of water". These concepts now apply also to underground waters.

In all States of the United States where the appropriation doctrine had applied, underground waters flowing in definite channels were subject to appropriation and to registration requirements. This is so either because all waters were declared to be public and subject to appropriation, or because water flowing in definite channels had been made public. In the western part of the United States most States subjected percolating water also to prior appropriation, and a number of States had enacted separate underground water codes. Others simply extended their statutory provisions applying to surface waters to include ground waters.

E.   Water law systems in socialist countries

Under the Soviet system of law, water, as well as land, subsoil and forests are mainly the exclusive property of the entire people, i.e. of the State. This type of water legislation directs the water economy of the socialist States and, with the exception of some use for local and domestic purposes, grants the right to use water to associations, co-operatives or agencies responsible for sectoral water resources development, rather than to individuals.

According to this system, legislation is only one instrument of ground-water management. Other instruments are planning, economic incentives, application of penalties as a deterrent, technical aspects, and public participation. Water is considered fundamental to production processes. In view of its importance for all aspects of socio-economic life (water supply, agriculture, industry, transport, etc.), water must be protected quantitatively and qualitatively. This is generally proclaimed either in the Constitution or in some other fundamental law such as environment laws and water acts. More detailed provisions are contained in specific rules, provisions, instruction or other administrative acts.

Water resources (surface and subsurface waters) are considered as natural assets of society and thus constitute public property. Their management forms part of the general State-administered management of water resources. Their utilization will be administered by the State, taking into account the interests of the different users with due respect to the general interest of the society. Therefore, all water production and water use is subject to licensing in socialist countries. The corresponding legislation leaves only a narrow sector open for free practices taking care not to endanger the general interest of society. However, individual abstraction of ground water for unrestricted domestic use is allowed, free of government permit, for example, in Bulgaria, Poland and Romania.

F.   Appraisal of the existing situation

Existing legislation in most ECE countries, regardless of historical or socio-economic differences, tends to impose restrictions regarding use of ground water on the owners and/or users of land, wells or other abstraction works. In civil law countries, the two concepts of "ownership" and "right of use" are separate. The administration is vested with legal powers to exercise control over (a) the quantity and quality of ground water abstracted, whether public or private, as well as (b) the return of waters to the aquifer and its protection against pollution in general. Although the trend appears to be in favour of firmer government control over ground-water resources through public ownership of the resource and/or control over the right of use of the "private" water resources, there is still a problem posed by the dichotomy between rights of ownership and rights of use, as appearing in legislation, infringing on the effective management of ground-water resources.

While private property status, accorded to ground waters in most countries adhering to the civil law system, may still constitute a hindrance to direct State control over ground-water uses and users, it was argued that it should not necessarily be regarded as a serious constraint provided that legislature and government had authority to restrain in the public interest the rights which accrue from private ownership of ground waters. In such a process, any existing rights acquired through land ownership or other legal title could be protected; any modification introduced by new water legislation could provide for their identification as water-use rights in order to bring them under a government-controlled permit system. Various procedures have been followed to this end, such as (a) simple notification of the right with exemption from the requirements of the new procedures for obtaining permits; (b) acknowledgement of rights as prior rights in time; (c) nationalization of water; (d) recognition of rights on a par with new water-use rights, provided waters are utilized in a beneficial manner or according to plan; (e) abolition of rights with indemnization or compensation procedures.

At this juncture it should be recalled that the purpose of water legislation is to ensure, on the basis of water availability, both in terms of quantity and quality, the optimum use of water resources and their conservation, in order to satisfy present and future demands for every purpose. Such an aim may be best achieved by bringing under control existing and possibly future uses of water. While certain aspects of underground water control are similar to that of surface water, and may be dealt with in a comprehensive water act; even so, special provisions are necessary for ground water, either as separate legislation or as a chapter in a general act. In either case, it is desirable that regulations should be consistent.

General water legislation, or a basic water code or act, particularly with reference to ground water management, could include provisions with respect to the following:

(a) Ownership should be clearly defined in any water act or code, with respect to both surface water and ground water. The tendency is to declare all water resources State, crown, public or common property. In the absence of clear definitions, the right of the State to at least control and regulate such ownership is indispensable;

(b) Rights of use must be distinct from ownership rights. Individuals and corporations may establish only a legal right to use water. A legal enactment could frame rules with respect to the origins of this right, various purposes of utilization as well as criteria for their recognition, reallocation or readjustment;

(c) Existing rights may be recognized consistently with availability of water, beneficial use, competing claims on the same source of water, provided such existing rights are defined as rights of use and not ownership rights; future use rights should be controlled and administered by the State; and in this respect legal protection of existing and future users and rights to use should be assured;

(d) Priorities to use ground water should be kept flexible so as to satisfy present and future requirements, changing circumstances, provided that first priority be given to human and animal consumption;

(e) Governments should have the power to control ground-water abstraction and use in areas endangered by depletion of aquifers or quality degradation through reduction, suspension or modification of existing rights to use ground water as well as powers to control and prevent pollution of aquifers through suspension or modification of land-use rights.

The theoretical principles outlined above are intended only as an illustration of some of the basic considerations involved. Each country, depending on its particular circumstances, must determine the best means to reflect its ground-water policy in appropriate legislation. Some principles may be disregarded or combined with economic instruments and other managerial tools best suited for particular policy issues and their sequence of implementation.

II. LEGISLATION AND GROUND-WATER MANAGEMENT

A. Joint management of ground water and surface water

The varieties of hydrological situations involving ground water are many and complex. They include interactions between ground water and surface waters, between ground water and other natural resources, and between ground water and other elements of the environment. Generally, ground waters are interconnected with overlying surface waters (streams, lakes, or seas). The entire water table may fall as a consequence of decreased natural recharge, over-pumping or significant diversion of surface water with which an aquifer is in communication. During the dry season, flow from ground water "feeds" surface waters, and constitutes, along with marshes and swamps, a natural regulator of rivers and lakes.

Conversely, of considerable significance is the reverse situation, whereby lakes and rivers feed into ground water. This takes place either as a temporary phenomenon during high water or as a constant effect, for example, where a stream disappears into the ground and "reappears" far away. An aquifer so charged may contribute to a distant lake or different river.

The interrelationships between surface and ground waters are various, frequently pervasive and of great practical significance. The inflow to, and the outflow from, aquifers can be reduced or obstructed by human actions. Regulation is required to control these activities. Failing to protect the quantity and quality of surface water may have a deleterious effect on ground water and vice versa.

The perception that surface and underground waters are generally interconnected and that they are simply distinct phases of one and the same hydrological cycle is finding its way into legislative frameworks for water resources management in a number of countries. This modern management approach - the integrated management and joint use of surface and underground water resources - is variously reflected in legislation. In the United States, New Mexico has a complex system whereby ground-water appropriations are only permitted on condition that surface water rights are given up in proportion to the ground water pumped. According to legislation in Colorado (United States of America), prospective appropriators may separately or in combination (a) use wells as alternate points of diversion of surface rights; (b) provide a substitute supply of surface or underground water to a downstream appropriator senior in time; or (c) develop a plan to increase the supply of available waters. Other laws implicitly recognize the interconnection by forbidding or restricting the drilling of wells within a prescribed distance of a surface stream or dormant water body. Provisions to this effect are in force in Belgium, Spain (Canary Islands), Portugal and Italy.

Joint use of surface and underground water resources tends to arise in circumstances of high demand on existing sources of supply, or in pursuit of alternative options for increasing available supplies. There are two distinct technical aspects to this management approach, with corresponding legal and institutional implications. One is the co-ordinated use of wells and nearby surface waters in a given area. The other is the large-scale manipulation of ground water and surface waters through techniques and programmes of aquifer recharge; storage of surface run-off in underground reservoirs; release of ground waters to maintain required surface stream flows, or vice versa; and the mixing of waters of different origin and quality. All of the above have the potential for encroachment on established, legitimate water rights and use patterns. In addition, while studies have been made of ground-water depletion, pollution and interaction with surface water landward, little is known about legal issues associated with losses of ground water into the sea and sea-water intrusion into aquifers, as a consequence of, inter alia, ground-water abstraction.

Underground water bodies with limited or no direct recharge from precipitation, and little or no interconnection with surface water bodies are deemed to be "independent" reservoirs not connected with the hydrologic cycle. Different rules may be necessary to regulate this type of "non-renewable" water resource.

The man-induced recharge of aquifers may play an important role in the management and planning of ground-water resources, and may be subject to regulation to ensure protection of the delicate natural equilibrium under which ground water occurs in its natural quality conditions. Specific provisions concerning this balance are contained in the legislation of Belgium, whereby all artificial recharge is subject to a prior statement of intent being filed with the responsible government authority. The statement can impose terms and conditions designed to protect the affected aquifers and existing abstractions.

Aquifer recharge and water storage therein are both subject to permit requirements and other detailed regulatory provisions under Florida's (United States of America) 1972 Water Resources Act. The Act specifically empowers the regional water management districts to construct works for the storage of water in, or withdrawal of water from, an aquifer. It further emphasizes that storage or recharge water must be of a "compatible quality".

The legislation of France and Spain vests the Governments with responsibility and authority to plan the combined use of surface and underground water. Responsibility lies with the river basin agencies in France and with the central Government in Spain.

A more general mandate to plan the development, conservation and use of ground water - without specific reference to combined management of surface and underground water resources - is found in the legislation of the Netherlands, where such responsibilities lie with the provincial governments.

A question may arise as to whether to separate the policy for ground water from that for surface water, covering the same factors and principles. This will depend upon existing conditions. Integrated water resources policy covering both surface and ground water is considered preferable. In this connection, joint management of ground and surface water resources holds the promise for better distribution of water and greater efficiency in use, less waste or loss of water in transit from the source or intake down to the user, and less need for storage and distribution and, hence, lower capital investment. These advantages are perhaps greater for irrigated agriculture.

Consideration should be given, therefore, to adopting joint management techniques whenever the circumstances so warrant. With a view to minimizing the effect of such techniques on established water-use rights, joint management should be adopted at the planning stage and before a crisis situation has developed regarding depletion and misuse of a particular resource. In addition, with a view to attaining efficiency in joint management, the responsible government authority should be empowered to modify surface and ground-water rights, and to revoke an unused or badly used right. If surface and underground water resources are handled by separate government departments, close co-ordination between them is essential.

The implementation of joint management programmes calls for some flexibility regarding water rights and corresponding licensing systems. This is not always provided in the legislation of most countries, especially with respect to transfers of rights and their adaptation to changing circumstances by administrative fiat. This is a sensitive area of water resources law; yet, it is necessary to strike a balance between the security of water rights tenure and the requirements of advanced water-management techniques. In addition, joint use can be limited by separate legislative and administrative treatment of surface and underground waters. Whereas surface and underground waters are usually regulated in a similar manner in legislation in a number of ECE countries, administrative arrangements often do not follow suit.

B. Exploitation and use control

The necessity of conserving ground-water resources has led to the regulation of their exploration and exploitation. As a result, individuals and public and private enterprises which intend to sink a well or borehole must comply with requirements ranging from the filing of a statement of intent with government authorities to the obtaining of a prior government permit and attendant reporting requirements.

This general procedure is subject to a number of qualifications which vary from country to country and reflect the conditions peculiar to each of them. Such qualifications pertain to: (a) the purpose of the ground-water use envisaged, since, as was mentioned earlier, landowners generally enjoy unrestricted privileges to bore and abstract water from under their own land at least for their own household consumption; (b) the location of the abstraction, as restrictions may be in effect in designated "critical" areas only; (c) the depth of the bore or abstraction; (d) the volume of water abstracted; (e) the means employed to abstract the water; (f) the legal status of the ground-water users; (g) time of abstraction; (h) failure to use the water; (i) non-compliance with quality criteria.

The legislation of France presents an interesting combination of the above-mentioned qualifying factors: the abstraction of ground water by public users, i.e., local authorities, is always subject to government permit, irrespective of the location, volume or purpose of the abstraction. Private users may freely abstract ground water (a) for their own domestic consumption and (b) for non-domestic consumption, provided that the rate of abstraction does not exceed 8 cubic metres per hour. If the limit is exceeded, a prior declaration must be made to the government authority. A government permit is required with respect to all abstractions made in designated protected areas, and which exceed a given depth, irrespective of the volume or purpose of the abstraction. Such depths vary from protected area to protected area, and range from 5 to 80 metres.

In Belgium, a government permit is, in principle, required with respect to all abstractions of ground water anywhere in the country. However, (a) all abstractions for household purposes, (b) all abstractions made with non-mechanical means, (c) abstractions made for testing purposes, and (d) temporary abstractions made in connection with public or private works, provided they do not exceed 96 cubic metres per day, are exempt from permit requirements, on condition that the aquifer tapped is not an artesian aquifer. Requests for government permits are subject to a rather complex evaluation process according to the volume of the abstraction.

Similar principles apply in Luxembourg, where ground-water withdrawals made with non-mechanical means and not exceeding a depth of 20 metres are exempt from permit requirements otherwise applicable, on condition that the aquifer tapped is not an artesian aquifer. In the Netherlands, a government permit is required if the abstraction exceeds 10 cubic metres per hour, whether the water is used for domestic or irrigation purposes.

It should be noted that the functional relationship between the sinking of an exploration well and large-scale water abstraction from it is often reflected through legislation in one permit authorizing both. However, the granting of a permit for the construction of a well or for drilling does not always imply the right to withdraw and use water. In some countries, this right is subject to a specific permit, granted when test pumpings have been carried out; the permit sets conditions governing, inter alia, the quantity of water which may be withdrawn. It would be desirable to establish exploration or prospecting permits as distinct from water use permits; they could have a short-term duration. Since water quantity and quality may not yet be known, such permits would provide the required flexibility and caution before the granting of a water-use permit.

In this review of government administered regulatory controls over the exploration and exploitation of ground-water resources, separate mention should be made of spacing requirements for wells. Legal requirements to this effect are designed to protect established abstraction rights from encroachment by abstractions later in time and to protect ground-water resources from overdraft and eventual depletion. Well spacing requirements are laid down by legislation in Cyprus, Spain and Turkey.

In view of the superior quality of underground waters compared to surface waters, the legislation of some countries safeguards ground-water resources for priority use as drinking, domestic, and public water supply. This principle is manifest in a number of different ways. Under the legislation of the USSR, the use of underground waters of drinking quality for purposes other than drinking and public water supply is, as a rule, expressly prohibited. In Romania underground waters can, by law, be used for other purposes only if the drinking water supply requirements are not thereby affected. The German Democratic Republic and Czechoslovakia seek to achieve the same end by means of pricing mechanisms for water at the source, designed to penalize withdrawal of ground water for non-priority purposes. Under the legislation of the German Democratic Republic, appropriations of ground water for industrial use cost as much as 10 times the price levied if surface water was used instead. In Czechoslovakia, a surcharge would apply, bringing the charge payable for industrial withdrawals of underground water to about twice as much as the charge for comparable withdrawals of surface water. In addition to economic disincentives, the legislation of the above-mentioned countries foresees that the use of ground water for purposes where drinking water quality is not urgently required could be permissible only as an exception under an explicit permit from water management authorities.

The future importance of ground water will predictably depend less on its quantitative share for meeting total water demands than on its role as a preferred source of supply for uses demanding the highest standards of quality and safety. Therefore, efficient allocation of high-quality ground water to uses demanding high quality should be encouraged. This could be achieved by adopting appropriate regulatory restrictions on lower priority uses of ground water, or by adopting pricing mechanisms for ground-water withdrawal at the source which penalizes lower priority uses.

Legislators in a number of countries are reluctant, however, to include regulatory restrictions or adopt pricing mechanisms to act against lower-priority uses of ground water. This is so where, for example, the concept of internalizing social and environmental costs has not yet been fully accepted. It is particularly true where important aquifers are already polluted, e.g. contaminated by nitrates, heavy metals, chlorinated hydrocarbons or other toxic, persistent and bioaccumulative substances, or where aquifers are endangered by pollutants already in the soil and leaching slowly into ground water. Thus, there may no longer be a striking quality difference between surface and ground waters in a number of cases. Indeed there are aquifers which have already been rendered less suitable for direct human consumption. In view of these considerations, every effort should be made - supported by appropriate legislation - to conserve remaining ground-water resources in order to meet the demand for highest quality water. This will be done by efficiently protecting these resources against pollution and over-exploitation.

Allocation of water according to amount, purpose and timing should be centralized in a water administration, whether at the State, basin or, as a minimum, sub-basin levels in the case of large basins. Institutionalized, co-ordinated and obligatory co-operation is indispensable among all existing water licensing authorities. The water rights administration should be empowered to recognize existing rights, grant new water-use rights, and issue licences, permits, authorizations or concessions, consistent with availability, competitive applications and government policy regarding water.

The introduction of permit and economic disincentive systems for the use of ground waters signals a major shift from the traditional supply-management approach to a demand-management approach conducive to a more rational use of this resource. The existing situation still leaves room for improvement, despite the prevailing trend towards firmer government control over ground-water resource exploration, exploitation and use. This applies particularly, as regards: (a) more effective implementation and control; (b) the introduction of dual systems of permits, one for exploration, another for exploitation; (c) more detailed specifications and requirements in the exploration and exploitation permits; and (d) restrictions on ground-water allocation to water users.

In all the legislation analysed, no consideration appears to have been given to the problems posed by the large-scale use of heat pumps using ground water. The increasing application of such devices suggests that this particular use of ground-water resources will also need to be regulated and closely monitored.

In the foregoing, measures offering protection against pollution and over-exploitation of aquifers were stressed, emphasis should also be given to legal measures offering protection against harmful effects of ground water. Adequate drainage of aquifers could be provided for in legislation where waterlogging, salinization, silting or other harmful effects of stagnant ground water occurs. Nature and man may both be responsible for raising the level of ground-water tables, resulting in negative effects on structures, agriculture and the environment or even impaired health.

A specific man-made problem with respect to the water table may arise when ground-water use patterns change. This occurs especially in industry and mining. Sometimes water abstraction from a particular aquifer is terminated after long periods of exploitation. Rising water tables may conflict with land-use practices established earlier under low table conditions. Difficult legal questions with regard to responsibility and compensation in case of damage remain unsolved.

Another aspect that is seldom reflected in ground-water legislation pertains to subsidence of soil and subsoil and the corresponding disorder of structures. There, too, legal provisions formulated in advance could facilitate resolving disputes over responsibility and damage compensation.

C.   Pollution control

The need to control water pollution in general and to protect ground water in particular, (barely recognized only a few decades ago) has been sharply felt in recent years, especially in industrialized countries. Water-pollution control is now aimed at preserving the natural quality of surface and ground water and protecting the environment which depends on such water, thereby decreasing existing levels of water pollution while protecting public health and assuring supplies of drinking water in the future. More attention is being given to protection of aquifers and control against pollution of ground water. Whereas measures to protect rivers and streams against pollution backed by sound pollution-control legislation might be expected to have an effect eventually, the first results of such measures for ground-water protection would become apparent only in the medium or long term. This is so because of the slow-motion, slow-response character of ground water. Nor does this resource facilitate direct quality assessment.

Certain difficulties with ground-water pollution control legislation derive from the absence of a generally accepted definition of pollution. The term "pollution" has a relative connotation. Moreover, pollution is not a static quality, it may change over time and space. From a legal point of view, it may be said that pollution denotes those changes which produce detrimental and lasting effects on water quality, caused directly or indirectly by human action, and thus impinges on the rights or interests of water users and conflicts with environmental protection concerns. A definition of ground-water pollution should cover all forms of contamination and deterioration of the physical, biological and chemical quality of water, including not only discharges into it of solid, liquid or gaseous substances but any agent affecting the water's characteristics. Further difficulties in

formulating and enacting pollution control legislation may stem from the fact
that scientific proof cannot always be established for specific pollution
phenomena in aquifers. Furthermore, legislative provisions relating to
ground-water pollution are often scattered through a number of legal texts.
Sometimes the provisions are overly lenient and ineffective or else too severe
and hence not strictly applied by responsible authorities. Therefore,
effective legislation is urgently needed in many countries to combat pollution
hazards which may impair aquifers for long periods of time.

In analysing national legislation to protect ground-water resources
against pollution, it may be concluded that countries take one of the
following approaches:

- Basic anti-pollution legislation is kept to a minimum (a) on the
  assumption that technical methods employed for water protection and
  purification are constantly changing, and regulations that are too
  specific soon go out of date, or (b) because of internal jurisdiction
  questions owing to a federal structure. In such countries, therefore,
  national legislation is general in character and more concerned with
  establishing a uniform policy and a legal framework (loi cadre) leaving
  special or detailed questions to be dealt with under subsidiary
  legislation or by federal States;

- Legislation is detailed in order to cover the entire field of
  water-pollution control in extenso although it may still leave certain
  issues to be dealt with under subsidiary regulation promulgated pursuant
  to the main provisions;

- Consolidated codes are enacted to cover management of all natural
  resources (including the environment) or protection of the environment as
  such. In both cases, ground-water pollution control may be included.
  One of the main characteristics of anti-pollution legislation is that is
  not only covers pollution of water but also of air and soil as well.

Whatever the approach chosen, there are similarities in legislation
devised for dealing with problems of water-pollution abatement, prevention and
control. Countries have adopted one or more legislative measures, including
special permits and licences for ground-water abstraction; licensing of well
drillers and those responsible for ground-water treatment, supply and
recharge; exploration or prospecting licences as distinct from ground-water
abstraction authorization; licences to supply or use ground water and
licences to discharge into ground water; amendments and modifications of laws
to facilitate supervision of ground-water abstraction and use; metering of
ground-water abstraction; setting of priorities for ground-water use for
specific purposes; general planning giving emphasis to land use; stricter
measures to protect ground-water resources against pollution and overdraft;
pollution abatement at source; establishment of special ground-water
protection zones; and prohibition of waste water discharge into and waste
disposal on top of aquifers.

Permits and licences

Open-ended obligations may be imposed upon the users of land and of
natural resources, in general, to refrain from disposing of by-products
resulting from various activities in ways which may result in the pollution of

underground waters. (Such provisions may be found, for example, under the USSR's Fundamentals of Water Legislation and under the legislation of most socialist countries.)

The prior authorization system, widely adopted in national water-pollution control legislation, may take two major forms or be a combination of both. In countries where the administration of ground water is part of the overall management of water resources, prior authorization controls and regulates the abstraction and uses of various water resources. Both the quantitative and qualitative aspects are thereby taken into account. Such a permit may specify identical conditions for abstraction, use and discharge of both surface waters and ground waters.

Other countries have given higher priority to ground-water protection. Surface water abstractions and discharges may be controlled by permits while such activities are virtually prohibited by various kinds of legislation in the case of aquifers. The Federal Republic of Germany has demonstrated that it sets even higher standards for the protection of ground water than it does for surface water.

Specific legislative approaches to preventing or controlling pollution of ground water vary from outright prohibition of all waste disposal and of land-use activities likely to encroach on the quality of ground waters, to the regulation of such activities by means of government-administered permit or penalty systems. Some laws impose land-use restrictions in designated ground-water quality control areas. Examples can be found in the legislation of many ECE countries whereby it is forbidden to deposit near a well any substance which might endanger the quality of the ground water in and around it.

In a number of countries, permits contain additional provisions to regulate such aspects as responsibility for injury or harm caused to third parties, obligation to purify waters after use and to install treatment facilities. Authorities try to exercise a certain amount of discretion in the granting of permits, which may be subject to revocation if the holder does not comply with permit conditions.

Permit systems range in scope from regulation of every operation which may have an impact on the quality of ground water, as in France and Austria, to the regulating of waste discharges of a specific origin. The latter, somewhat narrower approach is reflected in the legislation of the United Kingdom (England and Wales), Norway and Romania. In Hungary, as in many other countries, penalty systems are in effect whereby enterprises causing harmful pollution of waters - particularly underground waters - must pay a "pollution fine". In the United States concern for the quality of underground sources of domestic and municipal water supplies has resulted in the institution of a permit system for deep injection wells; furthermore, a legislative ban withholds federal funding for all projects which may result in contamination of designated "sole source aquifers".

Legislative measures for the protection of the quality of ground waters include government-administered regulatory controls which, although not directly aimed at protection of ground-water quality, nevertheless may make vital contributions towards it. In Malta, for instance, construction of new farmhouses and swimming-pools filled with sea water on main ground-water catchments is controlled. In France, pertinent restrictions stem from the legislation regulating classified facilities, construction of buildings, public health, and storage of gas.

In general such provisions may include directives relating to any activity which interferes with water quality, including the production, and use of chemical products and the disposal of urban, agricultural and industrial wastes. Provisions may comprise authorization or prohibition of construction and operation of (a) pipelines transporting oil, chemical products, coal slurry and the like, (b) storage and handling of solids, liquids and gases with a potential to endanger ground-water quality, particularly those which are toxic, persistent and bioaccumulative including PCBs, heavy metals, chlorinated hydrocarbons, mineral and tar oils as well as their products, organic compounds containing halogen, nitrogen and sulphur, to name only a few. It has been recognized that provisions should be preventive as far as possible and procedures should be established for ensuring co-ordination among responsible authorities.

In many countries of the ECE region, intensive agricultural development has introduced a new pollution factor: the widespread use of chemicals such as fertilizers and pesticides. Large-scale, intensive animal production and industrialized agricultural technologies have led to accumulations of wastes causing immediate or potential pollution, particularly of ground water. Several countries have recently taken legislative action to control this type of non-point source pollution.

In Finland, the construction of piggeries for more than 100 pigs is subject to the advance notification procedures already required for point sources pollution: information must be supplied about the size and type of manure containers and the fields where manure is to be spread. Only after an inquiry does the water authority approve the plan. In Czechoslovakia also similar regulations prevail.

The earlier "curative" approach still dominates much legislation, as opposed to the more advanced concept of pollution prevention. A more adequate and well-organized water-use permit system could greatly help in this direction. Under it detailed provisions would regulate the treatment of waste water or effluent discharge and especially the storage, handling and disposal of potentially hazardous wastes. Together with legal instruments, various other incentives (economic, technical and psychological) should be provided to abate existing pollution. The practice of issuing permits alone for discharging wastes into waters in a manner which is known from the outset to have deleterious effects is considered counter to the very rationale of the preventive approach.

Prevention of various kinds of pollution, such as that originating from point and non-point sources, from the the artificial recharge of aquifers, from the recycling of water and from accidental pollution, seems to call for more in-depth legal consideration. Penal and financial sanctions should be seriously considered along with other control measures.

Another threat to ground-water quality - growing in significance and not yet sufficiently considered by legislators - is created through the phenomenon of acid rain or, more precisely, acidic deposition, a term which includes both wet and dry deposits from the atmosphere. It can leach metals from the soil into ground water. This problem is intensified where acidic soils overlie slowly weathering granitic and porphyritic bedrock. Drinking water quality may be also influenced by the increased corrosion of reservoirs and piping systems owing to acidification of water. The degree and extent of the acid rain problem within the ECE region requires that effective legislative measures be taken in order to control emissions of sulphur and nitrogen compounds at least from major sources.

While relevant regulations have, of course, to be applied at the national level, it is well recognized that the problem has a region-wide dimension requiring effective international co-operation. Important steps in this direction have been already taken within the framework of implementation of the ECE Convention on Long-range Transboundary Air Pollution. To date some 20 ECE countries have committed themselves to reduce their national annual sulphur emissions or their transboundary fluxes by at least 30 per cent by 1983/95, using 1980 emission levels as a basis for the calculation of reductions.

Planning

Since pollution is often a consequence of water and/or land use - and affects both surface and ground water - the concept of integrated management of water resources has to be introduced in modern legislation for all types of water utilization and for the control of water quality and quantity. Ideally, from the early planning stage, legal provisions should allow for the assessment of possible impacts of potentially harmful human activity on aquifers, with regard to both quality and quantity. Environmental impact assessment procedures offer an approach which - backed by appropriate legislation - is a necessary component of ground-water protection during development activities. Ground-water monitoring should be established by law in order to keep its quality and quantity under continuous review in its natural state, before, during, and after human interventions.

As far as planning of ground-water use is concerned, many countries have enacted provisions in their relevant legislation which will lead to comprehensive strategies for long-term ground-water economy. Under these legal provisions numerous studies have been prepared, registers updated and forecasts established.

In Czechoslovakia, the German Democratic Republic, Hungary, the Union of Soviet Socialist Republics and other countries, the long-term planning of ground-water use is normally included in general schemes of water economy development elaborated for a period of 20 to 50 years. The plans are usually based on the principles of water law: for instance, on determining ground-water priorities in order to supply the population with drinking water. Similar long-term plans and individual perspective studies of water economy development covering ground-water use have been elaborated recently in France, Norway, the United States of America and other countries. In the USSR and other CMEA countries, a uniform methodology of ground-water assessment is applied in long-term planning. Four categories are evaluated:

- Category A, includes ground-water capacities examined in sufficient detail to explain fully the conditions for the use of these resources, their quality and content, the possibilities of their depletion and other aspects. These findings facilitate enumeration of costs for investments necessary for resource utilization;

- Category B includes reservoirs examined in a way that explains the main specific features of their deposition, composition and supply, as well as the water horizons and their relationship to other horizons and surface waters. These findings serve in the design of water abstraction installations;

- In category C_1 reserves are examined in a preliminary way in order to determine roughly the features, geological composition and deposition of water horizons. This knowledge is intended to facilitate assessment of the usable amount of water and help in determining the range of necessary exploration work; and

- Category C_2 assesses the ground-water resources according to knowledge of general geological-hydrological conditions. It serves in planning long-term water resources utilization and manpower distribution in territorial planning, as well as in the design of long-term water management measures.

From the experience gained, it may be said that planning procedures had focused in the past primarily on ground-water quantity assessment. Only recently have planning and forecasting of ground-water resources recognized qualitative aspects as well. It is becoming generally accepted that the objectives of planning need broadening thus serving not only the purpose of exploitation and use of ground-water resources but also, and in particular, the purpose of protecting aquifers against pollution and overdraft. Therefore, planning should among other elements, include quality-forecasts of ground-water resources for appropriate time horizons, taking into account pollutants from point and non-point sources of industrial and agricultural origin already in soil, and which would eventually contaminate ground water long after severe pollution control measures had become effective, if at all.

It appears that in many countries the quantity and quality aspects of ground-water management are still dealt with separately in different legal texts. The provisions so codified are then administered by separate institutions, often without apparent co-ordination. Legislation could, however, enforce institutionalized and obligatory consultation between, for example, the health services and those responsible for water resources management.

D.   Protection zones

An important practical legislative measure for the prevention of pollution of waters, in general, and of ground water or springs, in particular, is the establishment of protection zones. Within these zones certain activities are normally restricted if not prohibited entirely. Usually there are several zones, with restrictions becoming less stringent in zones more remote from the protected water. The legislative provisions for establishing protection zones also include rules designating their extent and empowering authorities to delimit such zones.

In most countries, it is fairly standard practice in legislation on ground-water resources to lay down special measures protecting vulnerable resources from depletion and/or contamination. Target areas may either be designated by the Government, when the criteria specified in pertinent legislation are met, or they may be identified directly by law with reference to standard descriptions (as in the case of the areas in and around sources of public water supply). Protection measures may range from restrictions on drilling of wells or ground-water withdrawals, to restrictions covering the entire range of land uses and land-use practices which may affect adversely the quality and the occurrence of ground water.

The level of complexity of pertinent regulations varies according to the scope of the legislation or the subject. Italian legislation, for instance, empowers the Government to designate areas where all exploratory borings are subject to a government permit. Similarly, in Belgium, the Government is empowered by legislation to designate ground-water quality protection areas and to subject all activities likely to contaminate ground waters therein to a government permit.

A greater complexity of approach is reflected in the legislation of Czechoslovakia, Cyprus, Finland, France, German Democratic Republic, Germany, Federal Republic of, Hungary, Netherlands, Switzerland. To a greater or lesser extent, the legislation of these countries envisages different classes of protected areas, with restrictions on land-use activities graduated according to the need for protection of each class and according to socio-economic considerations. The latter are of increasing significance as protection zones may cover relatively large surfaces: in Czechoslovakia, for example, the surface area of these zones represents about 10 per cent of the national territory while in Hungary it is about 7 per cent.

Swiss legislation provides for four separate protection zones, and restrictions for different kinds of land-use activities according to the relative importance of the underlying ground waters from the standpoint of provision to public water supplies. In Czechoslovakia, seven "protected regions of natural ground-water accumulation" have been designated pursuant to the 1973 Water Act. In the protected regions, a whole list of land-use activities is forbidden. In Finland and other countries, for example, certain prohibited land-use activities have been enumerated in appropriate ground-water legislation and/or subsidiary enactments. Ground-water protection zones for public water supply are divided into two different levels or degrees of protection, and land uses therein restricted accordingly. In the German Democratic Republic, the Water Law of 1982 envisages three kinds of protection zones, with corresponding kinds of land-use prohibitions and restrictions for the protection of underground sources of public water supplies. In addition, there are separate provisions aimed at preventing contamination of ground water from fertilizers applied in agricultural areas beyond the limit of protection zones.

Protection zones, especially interior protection zones near the places of ground-water intake are sometimes delineated in co-operation with water economy authorities implementing laws for water protection and the authorities of public hygiene under public health laws (e.g. Czechoslovakia, France). This is because very often water used for drinking or other domestic purposes as well as water used in food processing industries is regulated by public health legislation. In some countries, these provisions are included in special legislation dealing with water pollution, or in public health administrative provisions. Ground-water protection is supervised by water economy authorities in co-operation with the bodies of the hydrogeological service (USSR, Ukrainian SSR).

Since the establishment of protection zones may require expropriation of lands, provisions for compensation to landowners are usually included in this type of legislation. The purpose of such measures might be to protect a public water supply, or to prohibit the building of dwellings or other buildings if the resulting pollution could enter ground water or alter its character adversely.

The relationship between land and water resources takes on another dimension when ground-water reserves of critical importance, or occurring under critical natural equilibrium, need to be protected from depletion and/or contamination. Controlling the withdrawal of water and the disposal of waste in and around the source may not be sufficient. Restrictions on a range of land-use activities capable of adversely affecting the occurrence and quality of ground water may become necessary.

Provisions empowering the responsible government authority to designate "critical" ground-water protection areas - especially in and around the sources of public water-supply systems - and to restrict land-uses therein, are all but unknown to ground-water legislation in ECE countries. The implementation and enforcement of such restrictions, however, may entail dealing with delicate legal problems from the standpoint of protecting established, legitimate land-use patterns and relevant rights. Land-use planning has potential as a tool for preventing conflict and fait accompli situations in the protection of "critical" ground-water resources. The need for special measures of ground-water protection can be brought to bear on the land-use planning process.

In addition, where provisions are made regarding "critical" areas and relevant stand-by authority is given to the government, legislation may seek to strike a balance between that which is desirable from a strictly hydrogeological or hygienic standpoint and that which is socially and economically feasible under the circumstances. Equilibrium would be conducive to effective implementation of pertinent regulatory controls.

The legislator could also consider the need to restrict - on an ad hoc basis - water- and land-uses in case of accidental pollution or other emergencies impacting on ground water. Legislation may provide mechanisms for curtailing water abstractions and for restricting land-use activities in such circumstances. When sufficient power is given to the competent water administration immediate action can be taken in emergency cases, such as droughts, floods, depletion of aquifers, water rights conflicts, accidental pollution, intrusion of sea water, etc. The various tasks water management may have to cope with in emergency situations could be stipulated in law.

It must be particularly emphasized that the problems connected with ground-water pollution control require an adequate policy, as part of a global water-resources management policy. Water legislation, in general, and ground-water legislation, in particular, are the means whereby it is possible effectively to implement and enforce any desired water policy. Legislation in itself, however, does not constitute a panacea for solving problems. Any legislation, to be effective, must be the result of water policy decisions which should precede its enactment and be based on the political, technical, economic, social, legal and institutional factors prevailing in any one country. Moreover, water legislation is strongly influenced by the prevailing legal system. It takes into consideration the sociological, religious and philosophical character of the people of any particular country or region.

Water legislation is aimed essentially at ensuring, on the basis of water availability, sustainable use of the resource and its conservation, in order to satisfy present and future water demands for every type of utilization. This may be achieved by bringing under unified, co-ordinated or centralized administrative control the existing and future uses of water. A basic water act should contain also a statement of national policy on water-pollution control so that there is a clear-cut legal basis on which to promulgate subsequent detailed regulations. With particular reference to ground-water pollution control, legislation should attempt to achieve the following objectives:

(a) Be consistent with policies promoting: the rational use of water, prevention and control of water pollution, including transboundary pollution, water-demand management and integration and efficient application of suitable managerial tools;

(b) Strive for sustainable use of ground water while preserving, as far as possible, the natural quality of ground water; ground water should be recognized as a commodity with economic and ecological value;

(c) Promote preventive strategies and measures in harmony with socio-economic development policies; curative measures should nevertheless also be foreseen for emergencies, accidents and for aquifers already impaired in terms of quantity and/or quality;

(d) Allow for the elaboration of strategies that are flexible so as to respond expediently to changing water-use and land-use patterns, new legal, socio-economic and technical conditions, the more stringent demands on the environment, as well as growing public awareness and scientific knowledge;

(e) Ensure that either State dominion or State control be established over ground-water resources in order to allow monitoring and easier implementation of measures to combat pollution, particularly those which abate pollution at source;

(f) Protect aquifers as the proper unit for the management of the utilization and conservation of ground-water resources, especially by establishing and carefully monitoring protection zones;

(g) Interrelate protection strategies within overall water management and planning. Legal, administrative and regulatory measures should be co-ordinated with economic instruments and best available technologies;

(h) Pay equal attention to both quantity and quality aspects of ground water. Likewise, emphasis should be given to the joint management of ground water and surface water, while taking into account the distinguishing features of ground water as compared to surface water, which necessitate the application of special protective measures for aquifers.

The achievement of these purposes may be greatly facilitated by a consolidation of the legislative provisions governing water pollution, thus avoiding lack of co-ordination between measures, possible inadequacy in scope and lack of power and means to ensure effective action where and when required. In addition, an adequate (co-ordinated or centralized) institutional framework is also required.

III. IMPLEMENTATION OF LEGISLATION

Even if a good legal basis reflecting advanced ground-water policies exists, major difficulties would remain where there are no legal provisions for sound implementation of ground-water legislation together with all supporting measures necessary to promote sustainable development and rational use of this precious resource. Indeed, in most of the ECE countries one of the main difficulties lies in implementing the provisions of ground-water legislation.

A major hurdle for implementing water policies and legislation is created by the distribution of responsibilities among various institutions involved in water management. These institutions may be use-oriented (agriculture, industry, health, housing, environment, water, justice, etc.), or operate at different levels: international, national, regional, provincial, local or finally, governmental, para-governmental, or even private.

The rational management of ground-water resources calls for the administration of basic regulatory mechanisms in an integrated fashion. That is, it should be done by one government department or unit in charge of administering both quantity-oriented and quality-oriented control. Whenever administrative responsibilities are split along quantity/quality lines (or along other lines such as the end-uses of the resource) co-ordination among the government departments concerned must be ensured, lest the effectiveness of the management process be jeopardized. Legislation should therefore spell out in as much detail as possible the respective responsibilities of the various agencies as regards the implementation of water legislation.

It may be concluded from the survey that the institutional set-up for the implementation of ground-water legislation varies from country to country in the ECE region. The mandate covering ground waters may be specific, or it may be part of a broader mandate encompassing all water resources. Responsibilities may either be centralized, as in Hungary and Turkey, or they may be decentralized to the regional/basin level, as in France and the United Kingdom, for example, or to the local level as in, for example, Belgium and the Netherlands. Central and local (district) levels of administration may well coexist within a framework of a vertical or hierarchial type, as in Romania and Poland, or they may exist independently of each other.

In countries having a federal structure, where autonomous provinces or republics have substantial jurisdiction over water, problems may arise whenever ground-water resources lie under the territories of two or more such federated or autonomous units. The water legislation of Yugoslavia and Switzerland provides mechanisms to prevent inter-jurisdicitional conflicts in such cases. Yugoslav legislation lays down the principle that water resources planning and plan implementation are the responsibility of the federated republics. However, ground water of an inter-republican character cannot be interfered with from the standpoint of quantity or quality by one republic to the detriment of another without the latter's prior consent; a detailed implementation procedure is provided. The Federal Assembly enacted a Law on Inter-republic and Inter-state Waters in 1973 which was amended in 1976. By agreements between the Republics, they are subject to the water management schemes established by federal law. Under Swiss legislation, protection of the quality of inter-cantonal ground waters is the responsibility of the cantons concerned, but each must take into account the interest of the other(s). Apart from inter-cantonal agreements and court decisions, the Swiss Constitution now vests broad powers in the Federal Government to regulate water management. New article 24 bis specifically refers to ground-water resources and ground-water table regulation. It empowers the Federal Government to allocate water use rights among the cantons concerned whenever these fail to reach an agreement.

The Federal Republic of Germany reflects a federal system of a high degree of decentralization and as such it shows a trend to preserve and promote the initiative of the federal units (Länder) and simultaneously to ensure co-ordination at the federal level. The Länder have their own water laws. While on a national level, specific matters regarding ground-water protection are promulgated in framework regulations to which water laws of the Länder must conform, like other federations, the Federal Republic of Germany also encourages the member States to develop co-operation among themselves. The Länder have power to conclude water agreements among themselves or water treaties even with foreign countries. The present constitutional and legal position is that if there is a dispute regarding ground water between two Länder within the Federal Republic of Germany, then article 93 (1) No.4 of the Basis Law (Grundgesetz) comes into operation to decide on it as a constitutional matter, thereby bringing it under th jurisdiction of Federal Constitutional Court. When the dispute concerns rights, powers and obligations flowing from such a treaty, then paragraph 50 (1) No.1 of Federal Administrative Court Act 1960 comes into play and such a dispute then falls under the jurisdiction of the Federal Administrative Court as it concerns administration of a law as distinct from a constitutional matter.

In the United States of America, three mechanisms exist, namely (a) inter-State agreements; (b) court decisions through the operation of the jurisdiction of the Supreme Court; and (c) the exercise of a paramount federal power. Some 35 inter-State compacts (agreements) relating to water resources management and including, inter alia, matters of ground-water protection have been approved by the United States Congress. The general purpose of these agreements has been to achieve an equitable apportionment of water resources. Some of these compacts envisage a commission with powers to settle disputes arising out of the operation of the said compact. As there is no unified water law in the United States, the individual States have jurisidiction to pass decrees relating to the ownership and use of ground water found in their territory in so far as the federal Government is not competent. In the exercise of its powers to decide water disputes between federal units, the United States Supreme Court has pronounced many decisions of which some pertain to ground-water protection.

In federal countries, or in countries with units of governments having jurisdiction over water resources, comprehensive laws for ground water or water resources are often enacted at the federal level, setting out main policy directives and frames of reference as guidelines for subsequent adoption by component states or other units of government.

The rapid evolution of water management policies has led many ECE countries in recent years to review critically their water organization, including ground-water administration. The trend is now towards a more centralized or co-ordinated administration and towards integrated and joint management of surface and ground-water resources. To this end it seems advisable to conduct a policy of gradual consolidation of agencies and

functions of water administrations, with due regard to the conditions of each
country, in order to minimize one-sided economic interest. This facilitates
water resources planning and economizes on the costs of water institutions.
Likewise, it would be possible to carry out the administrative and supporting
measures stipulated by legislation and to implement smoothly any water
resources policy in a more consistent manner.

As regards federal States, the potential for constraint is directly
proportional to the kind and extent of powers over water which the federated
States or republics enjoy vis-á-vis the federal government. To the extent
that mechanisms for co-operation and co-ordination of measures and programmes
on both sides of an inter-State or inter-republic border are absent, the
management process is bound to suffer. In this connection, it might be
necessary to envisage the possibility that federal water administrations be
conferred more extended powers in the management of water resources,
particularly as regards planning and monitoring of activities in basins and
aquifers shared by two or more federated States. This could also cover
inter-basin transfer of ground water from one State to another. Allocation of
financial resources for water plans and projects would thus be facilitated as
well. This process is slowly taking shape in a variety of ways in the Federal
Republic of Germany, Switzerland, the USSR and the United States of America.

The problem relating to institutions and organizations for ground-water
conservation and development is a complex one, as the value of such resources,
like the value of all water in general, is closely linked to its use.
Furthermore, institutions and organizations for ground-water management should
not be separated from institutions and organizations for water resources in
general, since they are a part of the same resource. Therefore, ground-water
administration must be envisaged within the overall administration of water
resources.

Pollution being, in general, a consequence of man's activities affecting
water (both surface and underground), it is through the control of these
activities that pollution control may be exercised. The question is to see
whether a special pollution-control agency should be established or if such
control can be exercised by the authorities responsible for water management.
While a solution depends on local circumstances, it is not advisable to
multiply agencies administering water resources. Ground-water pollution
control may be better managed by the existing water-resources management
administration, in close co-operation with health and environmental
authorities and administrations concerned at different levels (national,
regional, provincial, local). Anti-pollution legislation should spell out
clearly the respective responsibilities of the various administrations
involved and confer the necessary powers on each one of them, including the
setting up of the mechanism for ensuring the implementation of anti-pollution
legislative provisions for ground-water.

Planning the conservation and development of ground-water resources
requires a comprehensive and unified approach. This, in turn, calls for
bringing water under efficient administrative management through the issuing
of water use or discharge permits, licences, authorizations and concessions,
as well as co-ordinating various control activities. Institutions and

organizations may be envisaged, whether at different levels - national, regional, basin-wide, local, international - or according to the function they may perform - policy-making, technical, executive, co-ordinating, advisory, administrative, legal.

The type of water organization, its nature, powers and functions, may vary from country to country, according to circumstances, and there is no clear-cut formula valid for all cases. The various issues raised may only serve as guidelines for consideration and study.

At the national level appropriate institutions are required to deal with political and technical co-ordination as well as with administrative control over water resources. This may be achieved by setting up at least a national water council composed of those ministers having sectoral responsibility for water. Such a body would ensure co-ordination at the highest political level, and would have the functions of (a) framing the overall water policy of the country; (b) deciding on the allocation of funds and water for different purposes; (c) deciding on reimbursement policies, payment of water charges and related financial policy matters, and (d) formulating policies with regard to water conservation and pollution control.

Parallel to this national council acting on a political level, there may be a technical and economic water board or commission composed of senior scientists, engineers and economists responsible in the various ministries and authorities including representatives of water users' associations. This body assures an institutionalized and obligatory inter-ministerial co-ordination of technical and ecomomic functions at the national level; it may be either advisory or executive.

At the national level a central water administration is necessary in order to: (a) execute in its name, and on its behalf, the political and technical decisions taken by the national council or commission; (b) administer the rights to use water through the issuing of water permits, authorizations and concessions; (c) evaluate and co-ordinate different projects before execution; (d) standardize and pool all information and data relating to water resources; (e) prepare a master water plan and regional plans; (f) control, authorize or execute individual projects; and (g) manage activities relating to ground water, including pollution control.

Regional institutions may be envisaged either as branches or departments of a unified water administration or as more or less autonomous bodies at the basin, sub-basin or local levels which are responsible for a particular water utilization project or area. In some countries, basins may be co-terminus with the size of the country itself. But, with the growing experience of inter-basin transfers, this form or organization may not always be adequate. In such cases, centralized and/or unified control at the national level may be more appropriate.

Local, departmental, district or municipal water authorities often exist at the lower level. Their activities need to be integrated or at least co-ordinated at the basin or sub-basin level, in order to avoid segmenting watersheds or aquifers by artificial administrative divisions. Water users' associations exist in many countries particularly where water utilization started through the activities of individuals. They are generally efficient, but sometimes slow in adapting to increased governmental control. Their creation often provides a solution to conflicting customs and traditional water rights, particularly with regard to ground-water uses.

In may countries regular courts or special water courts have jurisdiction over the adjudication of water disputes. With the centralization of water management in the hands of the State, administrative powers take over these judicial responsibilities, thus facilitating decisions on water questions between users themselves and between them and the water administration. More stringent establishment and application of clearly defined sanctions and punishments for those who violate water laws or the permit specifications is another element facilitating the implementation of, and respect for, the law. Due regard should be paid in designing sanctions to the degree of social danger of the violation.

Italy is one of those few countries which has special water tribunals or courts. These Regional Water Courts function at the regional level. There is also a Supreme Water Court. The Regional Water Courts have "first instance jurisdictional competences in the declaration of public waters, the delimitation of watercourses, lakes, river banks, and beds; matters affecting existing public water-use rights; expropriations in connection with the construction, operation and maintenance of public water and land reclamation works; the setting of compensation for damages resulting from public waterworks; and in cases of exclusive fishing rights, expropriation, and in water policy matters". In addition these courts are competent also "in the case of disputes concerning non-public ground-water resources provided such waters occur within a protected district and the responsible public water administration is concerned with such disputes". The Supreme Water Court usually functions as an appellate authority vis-à-vis the Regional Water Courts' decisions. All matters pertaining to customary water rights are settled by Regional Commissioners appointed for this purpose by Decree of the President of the Republic and vested with the required judicial powers.

There are no special water tribunals or courts in France and as such regular courts of law deal with litigation arising out of the implementation of legislative norms and the prosecution of related offences. However, according to the terms of the Law of 18 December 1968 and Decrees No.73-682 and No.73-683 of 13 July 1973 with the Administrative Tribunal Code, "Administrative Tribunals" have been established to deal with all contentious matters arising out of the implementation of decrees of the Council of State, interministerial and ministerial orders and administrative circulars which constitute what is known in France as, "Administrative Law". At present there are 25 such "Administrative Tribunals" in metropolitan France. They function under the control of the Council of State, which is the highest administrative jurisdictional authority.

In Spain the competence of ordinary law courts extends to matters relating to all types of water ownership and possession rights on beaches, river beds and banks, to corresponding easements and servitudes, to fishing, and aquatic birds hunting rights, to matters of preferential rain-water use rights; for matters of priority based on private law entitlements with respect to all waters not contained within their natural bed; and to matters involving damages and prejudices resulting from the sinking of ordinary and artesian wells, norias and underground galleries. The water legislation also provides for administrative tribunals competent to hear appeals against the

decisions of the administrative authorities in the water resources' field;
this pertains to decisions which prejudice the rights acquired by virtue of
corresponding administrative norms or which concern the imposition of
limitations or liens on private property in accordance with the law.

In Turkey, there are no water tribunals or courts and, as such, water
disputes are normally settled by the regular courts. The competent government
departments such as the General Directorate of State Hydraulic Works
(G.D.S.H.W.) and the Ministries of Power and Natural Resources and of Health
and Social Assistance have been given, through special legislation, some
judicial prerogatives. Their decisions are, nevertheless, subject to appeal
before the courts, including the Supreme Court, in cases of substantive law,
and the Council of State, in cases of administrative matters.

In the Soviet Union all disputes as to the interpretation of the
agreements between the administration and the collective farms, for example,
are resolved by the People's Court. If land is transferred to another
organization, the rights under the agreement are transferred with the land.
The protection of water rights as well as the correct use of land is covered
by the Criminal and Civil Law. The local Executive Committee can impose
administrative penalties for violation of local rules on the use of water,
which must be in accordance with the laws of the Republic. The criminal codes
of certain Union Republics provide penalties for offences such as unauthorized
use of water. Irrespective of the criminal law, civil law also provides for
liability against another party causing loss to a person entitled to the use
of water. It also covers protection of water works.

In the United Kingdom provision is made according to Water Resources
Act 1963, Section 116 (1), for the Secretary of State for the Environment to
establish, by order, a tribunal to which cases or classes of cases as may be
specified or determined in the order and for which appeals, normally made to
the Secretary of State, may be referred. Other special courts such as the
Land Tribunal, the Transport Tribunal and the High Court, as appellate
authority, are competent to deal with disputes relating to water rights.

Apart from institutional reasons, ground-water legislation has proved to
be ineffective in a number of cases, for the following reasons:

(a) Land users, whether it is for domestic, industrial or agricultural
purposes, still consider it their right to tap ground water underlying their
land uncontrolled and free-of-charge, as well as to dump any matter thereon
regardless of its potential impact on aquifers beneath;

(b) Control of abstractions from, and discharge into aquifers regarding
both quantity and quality, as well as control over land-use practices
impairing ground-water quality and quantity, is difficult and rather costly if
water and land users do not execute reliable self-control;

(c) Often relevant legislation does not permit easy access onto private
or State property by the personnel who normally would be in charge of
enforcement; and

(d) Lack of financial and human resources to implement ground-water
policies in an expedient and efficient manner.

Other constraints to rational management of ground-water resources include (a) reluctance to alter traditional views and practices of water as being abundant and free, (b) reluctance of water users to share an asset with their neighbours and with future generations, (c) lack of adequate data on quantity and quality, (d) difficulty of improving production processes to reduce water needs and water pollution, (e) slow adoption of new and more efficient methods of recycling and reusing waters, thus sparing ground water for uses which demand high quality, (f) slight interest in recharging of aquifers, (g) increasing waste disposal and insufficient waste management, and (h) increased concentrations of fertilizers and pesticides in agriculture. Special legal provisions are needed in order to deal with these problems. Ancillary measures might also facilitate the implementation of water legislation as indicated below.

An important means of implementing legislation would be to educate and motivate the public to adhere to the principles of rational use and protection of ground-water resources against pollution and over-exploitation. Public participation in ground-water management could be enlarged. While public education and participation is generally desirable in all aspects of economic life, in the case of ground water it is indispensable. The low level of knowledge as regards the nature, behaviour and vulnerability of ground water makes this imperative. It is also essential to achieve public acceptance of legal measures - restricting in general the unbridled freedom of individuals - to avoid resistance or outright opposition to the implementation process. Without such education and participation, any legal measure tending to control ground-water quantity and quality will be difficult to implement.

Implementation is also facilitated by the introduction of a more efficient form of supervision and control, both at the preventive stage, i.e. on the occasion of the issuance of water permits (to explore, abstract, divert, discharge or recharge) and after, i.e. on authorized activities, for the purpose of checking compliance, etc.

In order to achieve an effective system of water-pollution control, it is necessary to work out programmes of inspection of water abstraction, of water-purification and -supply facilities, and of field tests on ground waters to determine water quality. The water-pollution control legislation may provide the legal framework for the establishment of uniform systems of monitoring, methods of sampling, analysis of such samples, and data processing, thus eliminating problems of lack of uniformity throughout the country.

Legislation could stipulate that drillers and drilling firms be registered or licensed, and that they file a report to the competent authority for each well drilled. Drilling and sinking of wells should be carried out only by qualified persons licensed by Governments. They could also be obliged to give the authority prior notice of their intention to drill, sink or construct, enlarge, alter or simply repair a well. In this specific connection, there may be merit in extending the application of licensing requirements not only to professional well drillers but also to all who perform technical activities involving ground waters inside and outside an aquifer. In addition, the equipment used for ground-water abstraction, heat exchange, waste injection or ground-water recharge should also be subjected to licensing and continuous control.

Introduction of financial and economic instruments is an important element for ensuring rational management of water resources. The careful study, introduction and imposition of water subsidies, taxes, fees or even financial sanctions may constitute a disincentive for certain ground-water uses or incentive for protection measures. The relevant rules ought to be flexible in adapting to the local circumstances and time.

As far as protection of ground-water quality is concerned, the legislative measures are usually backed by economic tools similar to those applied to surface water pollution, e.g. fines. However, views on this issue are not consistent, especially in the case of large-scale or persisting ground-water contamination when the immediate polluter cannot be identified.

Many governments have enacted legislation which provides for financial provisions for the construction of municipal sewage and industrial waste-treatment plants. These financial provisions may be in the form of grants or subsidies (e.g. Italy, United States of America), low-interest loans, special bond issues tax advantages or government guaranteed loans.

Since water-pollution control measures may involve considerable expenditures, it is considered desirable that the principles governing the raising of revenue and the sharing of costs be established beforehand. A few countries provide for this in their water-pollution control acts (e.g. France, Finland), while others deal with matters of finance and revenue in special enactments.

Finally, the implementation of juridical provisions relating to water are closely connected with the solution of a number of scientific and technological problems. Scientific research is indispensable as a tool to facilitate the enactment and implementation of water legislation. The rate of change in the technology of water and ground water is very high, and it is easy for staff to fall behind. Training on the job and refresher courses are therefore helpful means to facilitate implementation and compliance.

Many countries have increased research efforts to shed light on transportation, fixation and leaching processes of pollutants in the subsoil, in both saturated and unsaturated zones, and even in deep-lying aquifers. Competent authorities and research institutes are responsible for this task. To give an example: in Turkey the General Directorate of State Hydraulic Works is authorized to conduct research studies on ground water as well as to make use of it. In order to monitor the pollution of ground water, several sampling centres have been determined and analyses are carried out by the department of drinking water in accordance with its annual work programme.

IV. INTERNATIONAL CO-OPERATION

It has to be admitted that international law on sustainable development, equitable use and joint protection of ground-water resources beneath frontiers and international boundaries is less developed than international law pertaining to surface waters crossing or forming the boundary between countries. International law, as stated in Article 38 of the Statute of the International Court of Justice, is based on (a) international conventions; (b) international custom; (c) the general principles of law, and (d) judicial decisions and the teachings of the most highly qualified publicists; to which should be added relevant activities of international organizations and the "final acts" of international conferences.

A.   International agreements

Recent treaty practice in the ECE region shows that some attention has been paid to ground-water resources shared by two or more ECE countries. The list, which does not purport to be exhaustive, includes the 1947 Peace Treaty between the Allies and Italy containing mutual guarantees given by Italy and Yugoslavia concerning the use of springs in the city of Groizia and vicinity; the Agreement between Yugoslavia and Bulgaria on water-economy questions of 1958; the Yugoslav-Hungarian Agreement with Statutes of the Water Economy Commission of 1955; the agreement between Poland and the USSR concerning the use of frontier water resources, of 1964; and a similar agreement on the same subject between Czechoslovakia and Poland, of 1958. While ground-water in these treaties tends to be a secondary issue, in 1967 there was a specific agreement on ground-water between Poland and the German Democratic Republic. Though established transboundary water agreements do not always mention explicitly ground water as an element in the treaty terms, a number of countries, such as for example Hungary, favour the interpretation that ground-water resources - being an integral part of the total transboundary waters - fall also under the jurisdiction of these treaties.

In August 1977 the Canada-United States International Joint Commission was asked by the two Governments to look into a question of transboundary ground water for the first time since the Commission's inception in 1909. Although the 1909 Canada-United States Boundary Water Treaty does not explicitly mention ground waters, article 9 of the Treaty is generally understood to include them by implication.

More recent treaty and negotiating practice shows specific concern for ground-water problems, also in conjunction with surface waters. An agreement between Upper Savoy (France) and the Canton of Geneva (Switzerland), concluded in 1977, lays down detailed rules for the controlled exploitation and the protection from pollution of the Lake Geneva aquifer, and provides for a programme of artificial recharge of the aquifer.

A draft agreement between Spain and France concerning the allocation of water in the Err River between the Spanish gore of Llivia and the French Legré valley reflects awareness of a link between surface water abstractions and ground-water levels.

In 1971, France and Belgium reached an understanding on curtailing withdrawals of ground water from a shared aquifer on both sides of the frontier. Furthermore, both parties recognized the polluting impact of the border river Espierre on this aquifer and are striving towards a solution.

Most of the above-mentioned agreements provide for the establishment of mixed or joint commissions of representatives of countries party to the respective agreement. The 1972 Convention between Italy and Switzerland on water pollution control, for instance, established a mixed commission to investigate the origins, nature and magnitude of pollution of surface and ground water which may contribute to pollution of Lake Maggiore, Lake Lugano and other waters.

Relevant to this topic are also two Directives out of many (which mainly deal with surface waters and quality protection) issued by the European Economic Community which are binding on its member States. One relates to the prohibition of pollution caused by certain dangerous substances discharged into the aquatic environment of the Community (4 May 1976). Another concerns the protection of ground water against pollution caused by certain dangerous substances (17 December 1979). Their ratification by the respective parliaments obliges EEC member States to adopt national legislation within a short time. National legislation may be stricter than the provisions stipulated by the relevant directives.

The treaty practice of ECE countries demonstrates an awareness of shared ground-water problems. However, in spite of agreements already concluded, many unsolved problems in the use of aquifers remain. Thus there is a need to intensify or expand intergovernmental co-operative arrangements in many ECE countries. For this purpose, the adoption of legally binding instruments between countries sharing the same underground aquifer is necessary.

Co-operative arrangements could include: (a) data collection, standardization and exchange; (b) establishment of a joint inventory; (c) research and training; (d) planning and management; (e) joint control and monitoring of activities with regard to quantitative and qualitative aspects of ground-water protection; (f) establishment of adjacent (common) protection zones; (g) establishment of commonly agreed land-use planning procedures and practices; (h) monitoring of surface and ground-water resources' behaviour and their interdependence; and (i) obligation to give notification concerning any activity which might modify the volume and/or the quality of water.

For such co-operation to work the establishment of a joint or mixed administrative mechanism is a prerequisite. Such machinery should preferably be set up for the management of both surface and underground shared water resources. However, in the case of ground water, such an institution would have to be in a position to control the activities inside the territories of the countries concerned. This could be handled through attributing adequate and clearly defined power to a joint or mixed institution.

B. General principles for co-operation

In the implementation of such co-operation, consideration should be given to a number of international declarations, resolutions and principles included in the:

  - Helsinki Rules, adopted by the International Law Association (1966);

- Relevant resolutions adopted by the United Nations General Assembly,

- Stockholm Declaration, adopted by the United Nations Conference on Human Environment (1972);

- Mar del Plata Action Plan, adopted by the United Nations Water Conference (1977); and

- UNEP Draft Principles of Conduct in the Field of the Environment in the Guidance of States in the Conservation and Harmonious Utilization of Natural Resources Shared by Two or More States (1978).

Indeed principle 21 of the Stockholm Declaration stipulates that: "... States have ... the responsibility to ensure that activities within their jurisdiction or control do not cause damage to the environment of other States or areas beyond the limits of national jurisdiction". Similarly, the Final Act of the Conference on Security and Co-operation in Europe (Helsinki, 1975) stipulates that: "the protection and improvement of the environment as well as the protection of nature and rational utilization of its resources in the interests of present and future generations, is one of the tasks of major importance to the well-being of peoples and economic development of all countries".

Pursuant to article 3 of United Nations General Assembly resolution 3281 (XXIX), each State in the exploitation of natural resources shared by two or more countries "must co-operate on the basis of a system of information and prior consultation in order to achieve optimum use of such resources without causing damage to the legitimate interest of others".

The need for progressive development and codification of the rules of international law regulating the development and use of shared water resources had been a growing concern of many Governments participating at the United Nations Water Conference in 1977. Therefore, the Mar del Plata Action Plan adopted at that Conference emphasized the concerted and sustained effort needed to strengthen international water law as a means of placing co-operation among States on a firmer basis.

The international practice of ECE countries provides a certain number of examples of ground-water regulation. These examples are nevertheless scattered and do not permit principles of international law specific to joint ground-water management to be deduced. In relevant declarations and resolutions of intergovernmental organizations, references to ground water are too scanty or vaguely implied in principles covering water resources or natural resources in general. Moreover, international courts do not appear to have rendered any decision specifically on ground water.

On a regional level, the Economic Commission for Europe by its decision D (XXXVII) on International Co-operation on Shared Water Resources "recognized the growing significance of an economic environmental and physical interrelationship between ECE countries, in particular where streams or lakes and related ground-water aquifers cross or are located on international boundaries". ECE thus called upon member Governments "to pursue and, if necessary, to strengthen their efforts to co-operate in the elaboration of policy aims, programmes and planning regarding the development, use and conservation of shared water resources". It further "encouraged member

Governments to continue their efforts to extend already existing international arrangements in the light of changing socio-economic requirements or of changing priorities in the utilization of shared water resources". Obviously these resources include ground-water aquifers where they are related to surface waters.

One conclusion may therefore be deduced whenever water resources within one hydrological management unit of common interest to two or more countries exist, it seems reasonable to assimilate the legal régime of underground water to that of surface water on the basis of a de facto connection between the two locations of the same resource. Since adoption of the Helsinki rules in 1966 by the International Law Association, the unity of a drainage basin in international relationships was considered as a desirable principle whenever it is not already a binding principle in the relationship between riparian countries.

On the basis of the Helsinki Rules, "Each basin State is entitled, within its territory, to a reasonable and equitable share in the beneficial uses of the waters of an international drainage basin" (Art.IV). What is a reasonable and equitable share is to be determined in the light of all the relevant factors in each particular case. These include:

"(a) The geography of the basin, including in particular the extent of the drainage area in the territory of each basin State;

(b) The hydrology of the basin, including in particular the contribution of water by each basin State;

(c) The climate affecting the basin;

(d) Past utilization of the waters of the basin, including in particular existing utilization;

(e) The economic and social needs of each basin State;

(f) The population dependent on the waters of the basin in each basin State;

(g) The comparative costs of alternative means of satisfying the economic and social needs of each basin State;

(h) The availability of other resources;

(i) The avoidance of unnecessary waste in the utilization of waters of the basin;

(k) The degree to which the needs of a basin State may be satisfied, without causing substantial injury to a co-basin State".

These rules set detailed substantive standards of State conduct in the development, conservation, use and administration of shared water resources. They are centred on two doctrines:

(a) The concept of the international drainage basin, as the aggregate of both surface and ground waters which, within a given geographical area, flow into a common terminus; and

(b) The concept of equitable use according to which each basin State is entitled, within its territory, to a reasonable and equitable share (to be determined on a case-by-case basis) in the beneficial use of the waters of the drainage basin.

The problem with international regulation of ground water lies in the fact that international law may lack the flexible, but specific principles needed to meet the exigencies peculiar to ground-water management, while great efforts had been made to develop legal principles for surface water. The majority of international treaties on water are limited to surface water problems but fail to encompass ground water. The result is that the legal principles followed for surface water have not yet passed - or maybe cannot pass - the test for sound ground-water management.

Despite relative inactivity in the field of international ground-water law in the past, international relations concerning ground-water resources are likely to develop rather fast, for two major reasons. First, the nature of the resource itself makes it an ideal subject for international co-operation. Second, countries are coming to attach increased importance to water, in general, and to ground water, in particular. Thus international co-operation may become increasingly vital. The greater desire of countries to protect this precious resource effectively against pollution and over-exploitation conflicting with a rapidly rising demand for its use will further induce riparian countries to negotiate.

V. FINAL APPRAISAL AND CONCLUSION

The diffficulties that ECE countries have noted in the formulation and implementation of national ground-water policies and legislation are many and varied. It is not possible to cite them all, as they vary according to technical, physical, economic, financial, social, legal and institutional factors particular to each country or area. Strong political will, education, training, knowledge and awareness of implications are necessary in overcoming these difficulties.

Where water resources are abundant and the impact of mankind negligible, management of ground-water resources may require no more than basic control mechanisms already available within the legislation of most ECE countries. However, the development, conservation and sustained use of ground-water resources under "critical" circumstances may call for management approaches - and corresponding legal and institutional mechanisms - of higer complexity. Reference may be made, among others to (a) joint use of underground and surface water resources; (b) protection of the availability of ground-water resources of a given quality by restricting land-use activities in designated ground-water-protection areas; (c) allocation of ground-water resources to uses requiring high-quality water; (d) control of ground-water abstraction and use, including in situ use; (e) control of waste water discharge above, into and near aquifers; (f) control of handling and disposal of potentially hazardous waste and material in general on the ground and underground; (g) control of aquifer recharge; (h) control of non-point source pollution.

While at present relevant legislation is being enacted often in response to "critical circumstances", it should be noted that conflicting situations regarding ground-water management are arising continually. As a consequence, relevant provisions and legislation should be enacted not only as "emergency" but as regular measures of ground-water management. In view of present ground-water developments, it may be safely assumed that this "critical situation" will arise in most ECE countries in a not-too-distant future reaching those which - at present - do not yet face critical situations.

As a follow-up activity it might be useful to examine managerial elements which should be taken into account in the readjustment, new formulation and implementation of legislation. Furthermore, water managers should identify the characteristics of ground-water legislation appropriate to the fulfilment of their tasks. Among these characteristics pertaining to modern ground-water legislation may be quoted flexibility, comprehensiveness, clarity, provisions for economic, social, environmental and administrative implications.

Although legislation is the means whereby it is possible to implement and enforce any desired water policy, it does not constitute, in itself, a panacea for solving all problems connected with water resources management. Any legislation, to be effective, must be founded on water policy framed according to the political, technical, economic, social, legal and institutional factors prevailing in the country. Water legislation, in turn, is strongly influenced by the general legal systems and must take into consideration the particular outlook and character of the people.

The task facing the countries of the ECE region is awesome. The challenge is to protect ground-water resources for continued prosperity now and for future generations. When an aquifer is impaired, quantitatively or qualitatively, its regeneration is extremely difficult or even impossible. People animals, plants - all need this vital resource now and in the future. Upon the drafters of laws, politicians and water managers devolves the duty to pursue vigorously any required action to preserve this resource.

ANNEX I

GROUND-WATER POLICIES AND STRATEGIES

By its hidden nature, below the surface of the earth, ground water is the most misunderstood of our water resources. While its quantity may be underestimated, its natural protection and self-purification capacity is often overestimated.

There are many misconceptions and even myths about ground water. Better knowledge of the properties and behaviour of ground water will not only help people learn more about this invaluable natural resource, but should also lead to better protection of ground water. Ground water can be considered as a global reservoir, from which a limited amount of water may be withdrawn without affecting adversely the quantity and quality of the water in the aquifer.

The role of ground-water supply in the total abstraction of water has been steadily increasing in many ECE countries over the past decades. A consequence of the intensive use of ground water, this trend is not likely to continue despite the rising standard of social welfare and further economic development. Humankind has always experienced problems regarding ground water, from both excesses and deficiences of this resource. However, in recent times serious concern has been expressed over the availability of ground water. For centuries people had assumed that ground-water sources were always pure. The massive use of chemicals has caused contamination of aquifers and, in several cases, deterioration of water quality beyond acceptable limits. Ground water is vulnerable. In order to avoid further irreparable damage, adequate protection and management have become priority issues.

The objective of ground-water management can be broadly described as ensuring that ground-water resources will be available in appropriate quantity and quality not only for the present but especially for future generations. With limited available resources of ground water, emphasis has to be given to water demand management. Priorities have thus to be allocated to competing uses, including drinking water supply, other water uses, and purposes other than water supply. In order to accommodate all the conflicting demands for, and activities within, ground-water reservoirs, comprehensive policies and strategies have to be introduced both at the local and national levels.

Protection of ground-water reservoirs has to encompass both quantity and quality. Quantity involves regulating ground-water withdrawals in such a way that a reservoir does not become permanently overdrawn and so that water supply sources do not influence each other (for instance pumping from closely spaced production wells must not lower the water table). Quality relates to controlling ground-water pollution (in any form) or countering it by effective and reasonable solutions to reduce pollution to a minimum. Control policies begin with the development of ways and means to prevent future pollution and to maintain the existing quality of ground water as pure as practicable. Strategies must take into account the distinctive hydrogeologic, socio-economic and environmental situation as well as the mechanisms for achieving the objectives set by strategies. Quantity and quality aspects are closely interrelated and cannot be considered independently. For instance, in lowering the ground-water table, the risk of inducing contamination of the aquifer, and thus impairing its quality, increases with infiltration of polluted surface water or sea water. Conversely, measures protecting against pollution would maintain the existing potential amount of ground water of a given quality. The search continues for solutions that would appeal to all interest groups affected by implementation of ground-water protection policies. A balance has to be found between what is theoretically desirable and that which is practically feasible.

Ground water is a major source of water in a number of ECE countries. Aquifers are particularly important for drinking water. Industries which are not supplied by the public sector also use considerable quantities of ground water, as does agriculture for irrigation purposes. Indeed, its extensive availability, in time and space, along with its generally excellent quality and easy exploitation, have made ground water an attractive and cheap source of supply. Recently, aquifers have been used with heat pumps, adding another and sometimes competing use to the classical ones. The various and often divergent demands on ground-water resources necessitate some form of arbitration by the responsible authorities. In many ECE countries aquifers have been considered as perennial reservoirs which could be tapped as the need arose, especially for drinking water. Conservation and protection measures appear to be major concerns. Adequate policies need to be adopted with a view to balancing short-term demands with long-term objectives, in the interest of present and future generations.

The close interaction between ground water and surface water has long been recognized. Not only does surface water have a potential impact on ground water, but the quality of aquifers has a direct impact upon the quality of rivers, both as sources of surface water supply and as terrestrial ecosystems. Interactions between rivers and aquifers are both natural (e.g. in karstic rock) or man-made (e.g. exploitation or artificial recharge of aquifers, bank-filtration). In view of this close relationship, it may be appropriate to harmonize, co-ordinate or combine the management of both categories of water resource.

Ground water is also linked with both soil and subsoil and, as such, it plays an important role in their mechanical and hydrological equilibrium. Damages to buildings and, more generally, effects on land-use patterns are well documented in many ECE countries and there are circumstances where ground water is causing complications that sometimes call for massive remedial efforts. This pertains, inter alia, to underground excavation. Mining and tunnelling are two typical examples. In certain cases, ground water causes geotechnical problems and is a muisance rather than a blessing. Here again, a co-ordinated approach to ground-water management and land-use planning is imperative and may call for arbitration procedures. Involvement of water managers at an early stage of the planning process is therefore essential (e.g. town-planning, siting of industries and waste disposal).

The quantitative and qualitative characteristics of aquifers not only depend upon the direct impact of their exploitation (abstraction or discharge), but are also related to various indirect impacts of activities affecting the overlying soil and subsoil. A number of examples may be cited:

- Construction of hydraulic works such as dams, canals and reservoirs can modify the dynamics of aquifers;

- Urban sprawl can deprive aquifers of part of their natural inflow by modifying the infiltration rate; conversely, leakage of distribution networks and sewage systems may spoil the quality of ground water;

- Pollution of surface waters may, in certain cases, contribute to contamination of aquifers;

- Agriculture impacts on ground water in various ways, both in quantitative and qualitative terms; irrigation and changes in vegetation cover (deforestation and afforestation) can influence the water table while the intensive use of fertilizers is likely to affect the quality of ground water;

- Effluent discharges as well as waste water treatment plants may become sources of local pollutation for aquifers;

- Storage or disposal of wastes, either on the surface in landfills or by injection into the subsoil, can have serious repercussions on the quality of ground water.

The vulnerability of aquifers to these impacts is subject to local variations, dependent upon the type of actions and the nature of the geological strata. Before undertaking any important project that could adversely affect the régime or the quality of ground water, water authorities sometimes conduct an impact assessment which allows for a comparison to be made between the objectives of ground-water protection and the value of the activities expected to cause disturbances.

Conservation and protection of ground-water quantity and quality are two closely related objectives. In those cases where alterations in quality arise from changes in the dynamics of an aquifer following intensive exploitation, the two aspects must be considered together. An example is salt-water encroachment. This is the shoreward movement of water from the sea into coastal aquifers as a consequence of ground-water abstraction. The responsible authorities may have to decide whether a limit should be set on the exploitation of ground water in order to preserve the quality of an aquifer, or to what extent artificial recharge could compensate for the intrusion of salt water.

It has been generally recognized that preventive measures to protect ground water are cheaper and more reliable than corrective actions. Prevention could include:

- Adoption of protective measures such as setting ground-water standards, regulations of various uses, ground-water zoning and land-use controls;

- Control of pollution sources by proper management and enforcement of regulations, identification of potential sources and evaluation of corresponding hazards, environmental impact assessment;

- Protection of abstraction points by determining protective zones for the intake areas and adequate design and operation of water wells.

Corrective actions could include soil removal after contamination, pumping barriers and interception trenches or putting an end to activities that are detrimental to ground water. In certain cases alternate water supplies may prove to be the only solution to problems of deteriorating ground-water quality. It should be borne in mind that ground-water reservoirs have a very long response-time and that corrective measures may not be fully effective because certain modifications in the water-soil interface are almost irreversible.

A wide variety of individual entities, including households, enterprises and local communities may influence the régime and the quality of aquifers. They may act directly as users or indirectly as producers or users of land and subsoil. These individual entities have their own particular objectives which are integrated into the socio-economic system of each country. For these users of ground water, the protection of aquifers is not a priori of prime concern, at least not in the short term. Water authorities have thus a role to play in ensuring an equitable distribution of rights and constraints among the different, and often competing demands. In view of the different local and areal conditions, and taking into account the specific natural (geographic, hydrological) and socio-economic situations, the objectives may be identified along with the ways and means to reach them.

Experience in different ECE countries has shown that despite the widely varying administrative and economic structures, public authorities have at their disposal only a limited array of possible actions to protect ground water. In most cases the objectives can be reached by indirect interventions through legislation and regulations, economic incentives, water demand-oriented management, land-use planning, and information and education. Protection of ground water will have to be integrated into a long-term water policy, adapted to the specific conditions of each country. This policy would aim at sustainable use of this vital resource which constitutes a commodity with economic value and involves rights subject to legal, social, and political constraints.

In order to reach the objectives mentioned above, several ways and means exist. Their implementation will have to be adapted to specific national and local circumstances as well as to the priorities defined by the authorities responsible for the overall management of water resources. However, there seems to be consensus on the general principles of ground-water management. According to the conclusions reached at the Seminar on Ground-water Protection Strategies and Practices, and in line with recent decisions adopted by the Economic Commission for Europe (ECE/WATER/38), the main ideas may be summarized as follows.

Ground-water management objectives cannot be formulated independently, but should form part of a general water policy which in itself is but one element in the overall physical planning process. Ultimately, ground-water management is closely related to socio-economic development and the welfare of the population. Thus management should aim at the highest possible level of protection for aquifers. Ground-water protection could best be achieved by ensuring coherence between the various objectives and in allocating an appropriate scale of priorities to them. In particular, security of supply of drinking water is an important element that has to be taken into account in formulating priorities.

Protection objectives by necessity differ, depending on local or areal circumstances and constraints; individual goals have to be defined by the competent authorities, taking into account the specific conditions. In each case, it would seem possible to delineate the limits of variations beyond which the régime and the qualitative characteristics of ground water become unacceptable. Such a definition of limits would correlate with quantitative and qualitative objectives for ground water. These objectives correspond to a compromise between diverging interests, both short term and long term. They would take into account the related impacts on surface water and could usefully be revised at appropriate intervals.

In implementing a comprehensive policy for ground-water protection, every user may be faced with the need to give up one or several independent actions for the common welfare. It may be difficult for an individual user to be concerned with the long-term advantages of preservation and conservation of aquifers when he or she cannot see how his or her own actions jeopardize the overall economy of an aquifer, much like the proverbial "drop in the bucket". Traditional attitudes may adversely affect the interests or actions of other users and by the time all users recognize the risks, remedial measures may be too late. Even if globally ground water is theoretically renewable, locally it may be limited in quantity or be vulnerable to irreversible damage. As the development of new ground-water reservoirs slows down competition for their use will intensify. Demand-oriented management, including arbitration and priority allocation will thus become a key element in ground-water policy.

In some sectors or during certain periods, conflicts of interests do occur and demands may be in excess of available supply. In such cases, appropriate policies promoting the rational use of water have to be implemented. In some instances, a policy of preferential allocation may be envisaged, giving appropriate weight to the competitive uses. The following criteria could tentatively be considered:

- Priority to drinking water supplies. This policy is currently applied in several ECE member countries and has been adopted by the Committee on Water Problems in its Declarations of 1980 and 1984 (E/1980/28 and E/1984/23);

- Preference to those sectors where ground water cannot be replaced by a competitive substitute, e.g. extensive irrigation in agriculture, certain industries;

- Preference given locally to a specific activity using surface water that is dependent upon an upstream aquifer which thus has to be protected (e.g. pisciculture);

- Economic value of an activity for which ground water is an essential factor or for which the withdrawal of ground water is a necessity (e.g. extractive industries).

Particular emphasis has been given to protection zones. These are used in order to protect aquifers for drinking water abstraction, especially in densely populated areas. Such protection zones seem generally advisable, even for future wells and in reserved aquifers. Protection zones play an essential role as preventive measures to protect aquifers at the local level. Their widespread application is to be encouraged. The following recommendations could be made in this respect;

- To adapt local easements to local conditions in order to avoid imposing unnecessary additional constraints which might lead to non-observation of the rules; for this, improved site inventories may be needed, with due recognition to activities that may be restricted by the easements;

- To increase, if necessary, the number and scope of the quality controls required for assessing conformity with prescribed easements.

It should be emphasized, however, that passive protection of ground water by means of protection zones is not sufficient in itself to ensure complete protection of aquifers. Indeed, the risk of accidental or diffuse pollution is not alleviated by such protection zones. A system of prior permits for activities which are potentially harmful coupled with strict control of their operation should form part of an overall approach to ground-water protection. Most ECE countries have already implemented licensing systems for major industrial sectors. Efficient co-ordination between active and passive methods would ensure the best results at minimal cost for society. It would avoid redundance between the respective protection measures upstream of a production site and downstream from a potential pollution source. In this respect, a central authority, responsible for overall water policy, has an advantage over a variety of different authorities responsible for various aspects of water management.

Legislation is generally recognized as an instrument of prime importance in the implementation of water policies, in general, and of ground-water management, in particular. Indeed, laws and decrees define the rights, duties and obligations of the various parties which have, either directly or indirectly, an impact on the quantity and the quality of ground-water resources. Furthermore, they spell out the rules by which specific activities such as abstractions, discharges and certain forms of land-use are governed. Through licensing mechanisms it is possible to ensure compatibility between human activities and protection objectives. Many countries have found that the efficiency of legal and regulatory instruments is considerably increased if the provisions of the laws and the legal status of ground water and surface water follow a coherent policy. In particular, it has been stated that ground-water resources, being an integral part of the natural water cycle, should have the same legal status as surface water which, in many countries, falls within the public domain. Equal importance should be given to the formulation of ground-water legislation and to its implementation.

Another element considered important for improving the efficiency of ground-water management is unified action by administrative units responsible for the implementation of the water laws and regulations. In this respect, entrusting a single water authority with the administrative management of all resources may be envisaged, without distinction among user sectors or between ground water and surface water. The territorial competence of water authorities should not necessarily be limited to either administrative boundaries or to water basins but should offer the possibility to encompass, as appropriate, specific adaptations to local conditions. Examples are aquifers common to two or more basins but which cannot be divided from a managerial point of view or deep aquifers that are independent of surface waters.

Administrative decentralization has been regarded as an appropriate means to improve the efficiency of water management. It is true that, generally, the transfer of competence to a lower level improves the communication and understanding between the public authorities and the micro-economic entities such as enterprises and individuals. However, it should also be underlined that local management has often a rather limited time horizon. Therefore the elaboration of long-term policies and objectives involving large areas might usefully be entrusted to more centralized levels of decision-making, at the basin or national level.

Legislation and regulations pertaining to water resources and their protection cannot be isolated from related activities involving land-use (land and subsoil). Harmonization of water laws, and especially of provisions covering ground water, with laws in other fields such as construction, town-planning, agriculture, industrial production, mining, transport and disposal of toxic wastes, seems essential. In particular, in areas where aquifers are unique and endangered, protection of ground water should carry decisive weight in land-use planning. In this respect, the identification of dumping sites of hazardous wastes and stringent control over sanitary landfill are critical elements. Co-ordination between authorities responsible for different aspects related to these problems (water, building, transport, planning, etc.) would seem desirable.

Apart from legislation and regulations, economic instruments are effective means to implement water policies and to ensure compliance with the objectives that have been defined. Financial incentives can be applied in different ways, through tariff policies, taxes or subsidies. They permit a transfer of financial charges compensating for any divergence between micro-economic entities and the aims of public authorities. Economic instruments could also contribute to a desirable allocation of ground water to specific users or preserve it for future use. If the charges levied contribute to the preservation or improvement of the aquifer, the financial transfers correspond in effect to internalization of costs. The same reasoning holds true for the application of the "polluter-pays" principle to contamination of ground water in those situations where the cost of rehabilitation is easy to evaluate. The question might, however, arise of whether this principle is generally applicable to ground water, as the sources of contamination, both in space and time, may be difficult to identify. With regard to quantity, one could ask whether it would be desirable or feasible to calculate the external benefits of ground-water exploitation accruing to others in cases where claims could be made when the corresponding advantages have ceased to exist. An example of such a problem is the rise of the water table after mining operations have stopped, causing flooding of basements in buildings or impacting on the stability and integrity of foundations.

It may be interesting to identify clearly the beneficiaries of ground-water protection measures, on the one hand, and the agents responsible for the change in the régime or quality of ground water, on the other hand. This exercise might prove useful in defining the modalities for the allocation of the costs of protection and for calculating compensation for damage. Better insight could be sought into the benefits and disadvantages, both direct and indirect, resulting from specific actions taken to protect ground water. Three main conditions have to be observed with a view to attaining the objectives of viable financial instruments:

- Appropriate co-ordination and integration with regulatory measures and their legal status;

- Compatibility with the overall economic conditions and principles prevailing in each country;

- Sufficient impact to constitute an incentive or disincentive.

The application of economic instruments should be in line with the general principles already adopted by the Economic Commission for Europe, and contained in recently adopted Declarations.

An important aspect of any policy is its acceptance by those who are involved, in whatever way, in its implementation. Active participation of all parties concerned in the choice of the objectives is essential for their recognition of the inherent obligations. The involvement of water users in the decision-making processes has been recommended already in the ECE Declaration of Policy on the Rational Use of Water: users can make significant contribution to the identification of targets and to implementation of protection programmes.

Any management decision and any application of specific legal, regulatory, technical or financial measures presuppose thorough and accurate knowledge of the general socio-economic context. Moreover, the policy-maker needs a meaningful analysis of the likely trends and impacts of any interventions.

Present knowledge of the mechanisms and phenomena inside aquifers, and the interface between ground water, soil, surface water and air, is sufficiently developed to provide the decision-maker with the basic tools for elaborating meaningful policies. Acute demands and increasing deterioration of ground-water resources in many ECE countries call for urgent action and effective implementation of these policies and strategies.

ANNEX II

GROUND-WATER LAWS AND RELATED LEGAL PROVISIONS
IN ECE COUNTRIES

This annex contains a country-by-country survey of present legislation and regulations in force at national and/or federal levels for the protection of ground-water resources against pollution and over-exploitation as well as for promoting rational use of ground water in the various economic sectors. This survey is intended to provide detailed information on existing ground-water legislation in ECE countries, thus illustrating the analysis presented in the main document. It was based on information contained in discussion papers transmitted by Governments to the ECE Seminar on Ground Water Protection Strategies and Practices (Athens, Greece, 1983) as well as on pertinent legislation of ECE countries, such as constitutions, civil codes, specialized water-resources' laws and other enactments.

Relevant information on national ground-water legislation was made available by the following countries: Austria, Belgium, Canada, Czechoslovakia, Finland, France, German Democratic Republic, Germany, Federal Republic of, Greece, Hungary, Italy, Luxembourg, Netherlands, Switzerland, Turkey, and United Kingdom.

Austria

Basic legal provisions regarding ground-water protection are contained in the Austrian Water Law (1959) and the Water Works Promotion Law. The Water Works Promotion Law distinguishes between projects of special importance for the country as a whole, which are therefore solely financed from federal funds, and other projects partly financed by federal authorities and partly by provincial authorities. At the level of the federal provinces (provincial governments) there also exist regulations on land-use planning and environmental control.

For the federal authorities, the decree on the elaboration and maintenance of a Cadastral Survey of the Water Economy (1969) constituted the first legal basis for the elaboration of material required for water management. Further important steps towards preparing the legal basis for the elaboration of fundamental material in the interest of long-term water management were the Law on Federal Ministries (1973), laying down co-ordinated planning with regard to water management, the Law on Federal Institutes (1974) ordering the Federal Institute for Water Balance in Karst Areas to make a survey of the water régime and water resources in karst areas. Finally, the Hydrography Law (1969) gave legal grounds to a considerable extension of ground-water observation networks and the further development of methods and procedures for working out and utilizing quantitative ground-water data.

The Austrian Water Law deals with the protection of water supply facilities. A distinction is made between protection areas, in which the erection of certain installations is prohibited, and partial protection areas, where certain activities are subject to authorization within the framework of the water law and where the management of some plots of land may be limited by the authorities. These measures may also serve to protect the future water supply. Partial protection areas are to protect ground water against impairment of its quality, i.e. all development liable to impair the quality of ground water over long stretches of its flow are subject to authorization.

Paragraphs 30 and 31 of the Water Law lay down the general obligation to keep water, including ground water, clean and free from pollution. In the view of Austrian experts, this obligation should apply to all areas containing potential ground-water resources suitable for use or development, whether or not they lie within protection areas or partial protection areas. This would provide basic protection of ground water against pollution by dangerous substances, i.e. substances which are not biodegradable.

Other pertinent legal provisions are connected with land-use planning, i.e. the allocation of land for various types of uses. The provisional laws on land-use planning provide, in principle, for the possibility of including objectives essential for water management within the framework of local land-use planning. This procedure is somewhat similar to the one connected with the establishment of a partial protection area, even though a land-use plan cannot comprise as many details concerning water management as an order for establishing a partial protection area. Nevertheless, a local land-use order can also make an important contribution to ground-water protection by the refusal to allocate plots of land for waste dumps or by limiting the allocation of land for industrial purposes.

Belgium

Provisions relating to the protection of ground water against over-pumping appear in several legislative texts, and, in particular, in the Royal Decree of 21 April 1976 regulating ground-water use, as amended by a Royal Decree of 5 June 1978. The most important provisions are quoted below:

"Article 1. Any new ground-water catchment within the meaning of article 1, paragraph 1, of the Decree-Law of 18 December 1946 instituting an inventory of ground-water reserves and establishing rules for their use, shall be subject to previous authorization.

A group of water catchments controlled by a single user and withdrawing water from the same aquifer shall be regarded as a single ground-water catchment.

Article 2. The following shall be exempt from authorization, provided the water does not constitute an artesian spring at the place of tapping:

(a) Ground-water catchments intended for the domestic needs of a family;

(b) Wells from which water is withdrawn by non-mechanical means;

(c) Test pumping operations lasting not longer than two months, carried out in order to ascertain the characteristics of an aquifer or to determine the characteristics of a future catchment, provided that the water withdrawn is neither used for an industrial purpose nor fed into a water distribution network;

(d) Ground-water pumping operations of a temporary nature carried out in connection with public or private construction or civil engineering works if the flow rate does not exceed 96 m^3 a day.

Article 3. The following shall be subject to prior authorization:

(a) Any change made to a ground-water catchment, subject to the provisions of article 2;

(b) Any transformations whereby a catchment ceases to meet the conditions set forth in article 2;

(c) The bringing into use of a ground-water catchment after a two-year period of continuous non-use.

Article 4. For the purposes of the authorization system, ground-water catchments are divided into two categories:

Category I comprises ground-water catchments with a daily flow not exceeding 96 m^3, as well as ground-water pumping operations of a temporary nature carried out in connection with public or private construction or civil engineering works if the flow rate exceeds 96 m^3/day.

Category II comprises ground-water catchments with a daily flow of over 96 m^3, other than pumping operations of a temporary nature that are covered by the preceding subparagraph."

The ensuing articles specify the authorization procedure for ground-water catchments in categories I and II. Every ground-water catchment must form the subject of a declaration which contains information concerning the applicant, the activities of the establishment, the purpose for which the water is to be used, the volume of water to be withdrawn, the place and nature of the catchment, the technical characteristics of the works, the nature of the withdrawal device and its maximum capacity, and the place of discharge of the waste water. The authorization has to mention:

"... the conditions to be fulfilled with respect, inter alia, to the maximum amount of water to be withdrawn per day, the withdrawal devices, the isolation of the various aquifers, the water-level records, the utilization of the drawn-off water, the preservation of ground-water catchments in the surrounding area, the preservation of properties on the land surface, and public safety."

The competent authority has to examine the applications and draw up a report.

"To that end:

1. It shall ensure, if necessary by means of an on-the-spot inquiry, that the draw-off of water applied for corresponds to the applicant's needs;

2.   It may require the applicant to submit a technical report showing that the aquifer can supply the requested flow without damage to nearby catchments, to property on the land surface, or to public safety;

3.   It shall transmit a copy of the application to the Geological Service, which shall report back to it;

4.   It shall consult the local authority of the place in which the catchment is situated; it may also consult any individual or legal person and any interested administration; 30 days after inviting them to state their opinions, it may proceed."

Article 15 provides that:

"Any new authorized water catchment shall be established in such a way that the water-level therein can be measured at any time, either by means of a sounding line or by an automatic device."

Article 21 bis provides that:

"Any attempt to recharge the aquifer shall form the subject of a prior application to the Director-General of Mines. The latter shall grant the application, possibly subject to conditions designed to preserve the integrity of the aquifer and of the catchments already in existence."

The Act regulating the operation of ground-water catchments (9 July 1976) defines the term "ground-water catchment" in the following way:

"The term ground-water catchment shall be understood to comprise all wells, collectors, drainage devices and, more generally, all works and installations whose purpose or result is to effect a ground-water extraction, including devices to tap springs at the point of emergence."

Articles 3, 5, 6 and 7 of the Act provide as follows:

"Article 3. The King shall appoint the officials responsible for supervising the execution of this Act and of the orders issued in implementation thereof. The King shall determine their powers of detecting and investigating violations which shall be without prejudice to the powers of the officers of the criminal police. The officials appointed under this Act shall have access at all times to the works and places of operation placed under their surveillance."

"Article 5. Violations of orders issued in implementation of this Act shall be punished by a fine of 500 to 2,500 francs.

Article 6. In the event of a repetition of the offence within 12 months from the date of the previous conviction, the fine provided under article 5 shall be doubled.

Moreover, the works in question may be immediately impounded and placed under seal.

Article 7. The heads of firms, landowners, users and operators shall be responsible under civil law for fines imposed upon their directors, managers or other agents."

The Royal Decree establishing provincial co-ordinating commissions on water problems (10 May 1967) lists in article 4 the commission's duties as follows:

"1. To exchange information on proposed hydraulic operations which are deemed necessary, especially those relating to land drainage, the scouring and improvement of watercourses, irrigation of agricultural lands, watercourse pollution and the supply of drinking water;

2. To compare the various work and research programmes and endeavour to achieve the requisite co-ordination among them and to that end, to suggest a scale of priorities for the various operations, as well as schedules for their execution;

3. To express opinions and make proposals with regard to the establishment, extension, merger or elimination of polders and drainage works in the province and with regard to the operation of the boards in question."

Canada

Within Canada's constitutional framework, the responsibilities for the management of water resources and for regulating their use are shared between the federal Government and the provincial governments. The Constitution Act (formerly the British North America (BNA) Act of 1867) provides for the distribution of legislative powers between the two senior levels of governments, and is the constitutional basis for water resources management. The Canada Water Act, proclaimed in 1970, provides the framework for joint federal-provincial management of Canada's water resources. This Act does not specifically refer to ground water; however, a definition of water resources implicitly includes ground water.

The provinces are generally regarded as the proprietors or owners of water resources within their boundaries; in each province or territory there is at least one agency responsible for the management of water resources. Responsibilities for ground water are usually divided between management for water quantity through the agencies' powers to approve and license water use, regulate flows and levy fees, and management for water quality through the agencies' powers to authorize development such as siting, construction and maintenance of septic tanks, waste-disposal sites, etc. Relevant examples are given below.

Well drilling: To maintain a measure of control over ground-water development and to provide for the collection of information on wells drilled, all provinces, with the exception of British Colombia and Newfoundland, have enacted ground-water or well-drilling legislation. Such legislation generally stipulates that all drillers or drilling firms must be registered or licensed and that they must file a standard report with the province for each well drilled. Most provincial agencies have established or are formulating legislation outlining procedures to be followed for the construction, maintenance and abandonment of wells along with restrictions on siting of wells with respect to buildings, power lines and sources of contamination. The water well reports, required by the provincial agencies, provide a vast body of information on ground-water occurrences in Canada (an estimated half million files).

Water Use Permit: Provincial legislation providing for the licensing of ground-water usage often recognizes "priority" of use. The drawing of water for private domestic and farm purposes is considered the most important use, generally followed by municipal water supply. Water for industrial, commercial and irrigation purposes is often regulated by the availability of the supply, the efficiency of use, the effects of the proposed withdrawals on existing uses, and established uses in the area. In most cases, domestic uses are exempt from licence requirements.

In resolving interference problems, priority in time and the adequacy of an affected water supply prior to interference are usually basic considerations. In principle, those holding the most recent water rights would be the first to lose them in a water-short situation, but in practice, governments are reluctant to use this approach. Instead, several users of the same aquifer may be asked to reduce their abstraction regardless of their priority in time. A variety of legislation exists; for example: in Alberta, under provision of the Water Resources Act, a person with a higher priority use, notwithstanding priority in time, may obtain a right to use water being used by a licensee with lower priority; in Manitoba, the use of ground water from aquifers containing high-quality water of limited capacity is restricted to high priority uses such as domestic and municipal.

Czechoslovakia

The basic provisions for legislative protection of ground water in the Czechoslovak Socialist Republic are contained in Federal Act No.138/1973 Sb on water (Water Act) and in Acts No.130/1977 Sb and No.135/1977 Sb passed by the Czech National Council and the Slovak National Council, respectively, covering State control of water management. Major provisions are given below.

Act No.138/1973 Sb on water (Water Act) stipulates particularly the following:

Article 1 - Purpose

Surface water and ground water, as one of the basic natural resources, forms an important part of the natural environment and serves to meet economic and other national needs. The purpose of the law is to protect bodies of water in every way, as they are irreplaceable and of primary importance to society.

Article 4 - Obligations when using bodies of water

Paragraph 1 - When making use of surface water and ground water, it is imperative to ensure its protection and to safeguard its economic and effective utilization.

Paragraph 3 - In the construction of housing developments, new industries and similar facilities, as well as in reconstruction, the investors are obliged to provide for their water supply as well as the treatment and discharge of waste water in a manner which is in no way detrimental to the quality of surface water and ground water.

Article 8 - Licence to handle water

A licence issued by the water management office has to be obtained for discharging waste water and effluents into surface water or ground water.

Article 18 - Protection of natural water catchment areas

Areas which, owing to their natural conditions, form an important natural water resource can be proclaimed by the Government of the Czech Socialist Republic or the Government of the Slovak Socialist Republic, respectively, as protected catchment areas. By doing so, the Government either regulates or bans activities affecting water conditions in such areas.

Article 19 - Protected zones

According to need, the water management authority sets up protected zones to protect the abundance, quality or purity of water resources. After consulting the organizations affected, the water management authority may ban or curtail the current use of properties or activities endangering the abundance, quality or purity of water resources.

Article 23 - Discharging waste water and effluents into surface or ground water bodies

Whoever discharges waste water or effluents into surface or ground-water bodies is obliged to ensure that the quality of the latter will not be impaired or in any way negatively affected. To this end, dischargers must assure, in particular, treatment of waters to be discharged in a manner in keeping with contemporary technologies available.

Article 25 - Protection of surface and ground water against pollution by substances other than waste water

Persons or organizations handling substances which are not waste water but can affect the quality or purity of surface and ground water must follow special regulations which specify the conditions for handling such substances, with a view to the protection of surface and ground waters against pollution.

Article 26 - Serious pollution

Persons or organizations causing or ascertaining a serious deterioration in water quality or danger to the quality of surface or ground waters are obliged to notify without delay the People's Committee (local authority) or the police.

Article 27 - Rectifying measures

Persons or organizations violating obligations concerned with the protection of surface and ground waters are charged, according to need, by the water management authority with measures necessary to rectify unsatisfactory conditions, in particular in order to prevent further pollution or threats to the quality of surface or ground water.

Article 29 - Compensation for loss of ground water

An organization which causes the loss of ground water or curtails the abundance of this resource, or impairs the quality of the said water, is obliged to compensate for the damage caused to persons or organizations granted permission to draw ground water from the given source.

Article 44 - Fees for discharging waste water into surface or ground water

Organizations discharging waste water into surface- or ground-water bodies are obliged to pay fees set by the Government of the Czechoslovak Socialist Republic.

Czech National Council Act No.130/1977 and Slovak National Council Act No.135/1977 Sb on State control in water management include the following provisions:

Article 10 - People's Committees supervise water management

In their territory, the People's Committees are responsible for continuous supervision, mainly by upholding their own decisions, of the provisions of the Water Management Act on the protection of water and its quality.

Article 11 - Supreme supervision of water management

The Ministry is responsible for supervising the subordinate water management authorities in implementing the Water Act and the regulations issued in accordance therewith.

Article 12 - Water management inspection

Water management inspection can be set up by the Ministry as an expert supervisory body in matters pertaining to the protection of water quality and water management. The water management inspection carries out tasks entrusted to it in connection with the supreme supervision of water management (cited under article 11) to an extent determined by the Ministry.

Article 16 - Water management officers of enterprises and corporations

Organizations drawing or otherwise using water or releasing waste water in quantities - or quality - exceeding the terms permitted by the Ministry in agreement with the central authorities concerned are obliged to employ trained personnel to ensure management of water and adequate treatment of waste waters.

Detailed stipulations concerning the protection and use of ground water are contained in the following regulations issued on the basis of the Water Act and the Act on State Control in Water Management.

- Czech Government Decree No.25/1975 Sb and Slovak Government
Decree No. 30/1975 Sb setting indices on permissible degrees of pollution,

- Czech Government Decree No.26/1975 Sb and Slovak Government
Decree No.31/1975 Sb on fines for infringing regulations concerning water
management,

- Czech Government Decree No.85/1981 Sb and Slovak Government
Decree No.46/1978 Sb on protected areas of natural water catchments,

- Regulations No.63/1975 Sb of the Ministry of Forestry and Water
Management of the Czech Socialist Republic and No.170/1975 Sb of the
respective ministry of the Slovak Socialist Republic on the duties of
organizations to provide information on ground-water resources and data
on the amount of abstraction,

- Regulation No.49/1975 and Regulation No.97/1975 Sb, respectively, issued
by the Czech and Slovak Ministries of Forestry and Water Management on
the procedure of water management authorities when changing current
practices to comply with the Water Act and the procedure when issuing
permits to discharge waste water into public sewers,

- Regulations No.42/1976 Sb and No.66/1976 Sb, respectively, issued by the
Czech and Slovak Ministries of Forestry and Water Management with regard
to water management officers,

- Regulations No.99/1976 Sb and No.158/1976 Sb, respectively, issued by the
Czech and Slovak Ministries of Forestry and Water Management concerning
water guards,

- Regulations No.6/1977 Sb and No.23/1977 Sb, respectively, issued by the
Ministries of Forestry and Water Management of the Czech and Slovak
Republics for the protection of water quality in surface- and
ground-water bodies,

- Directive No.51/1979 issued by the Ministry of Health of the Czech
Socialist Republic, "Public Health Regulations", on basic hygienic
principles for determining, defining and using protective areas of
ground-water resources designed for public drinking water supplies.

Finland

The main provisions on ground water are contained in the Water Act.
Points of law reflecting upon ground-water conservation are also found in the
Extractable Land Resources Act, the Building Code, the Public Health Act, and
legislation governing the use of toxic substances and pesticides.

The Water Act includes, inter alia, regulations prohibiting pollution of
ground water from depositing or conducting wastes, refuse, chemicals or other
hazardous substances so as to cause ground-water pollution on another party's
land. According to regulations, ground water must not be used in such a
manner nor should any steps be taken that would reduce the supply of household
water on another party's land. Nor should any use essentially prevent the
exploitation of abundant ground-water resources, or obstruct the water supply

to any plant using ground water. Ground-water pollution is also expressly
forbidden, but a water court can grant a permit in exception to legislation
forbidding the altering of ground water. The Water Act also regulates the
establishment of ground-water protection zones (see annex to document
WATER/SEM.10/R.18).

In accordance with the Water Act, a permit by a water court is required
for:

- Ground-water uses leading to consequences referred to in regulations
  forbidding the altering of ground water,

- Constructions and uses of ground-water intakes with a capacity of at
  least 250 m^3 of ground water per day, and

- The intake of ground water from another party's land, unless the owner of
  this estate has so agreed.

If the project concerned poses no greater risks, the water court normally
grants a permit for ground-water use, although not always for volumes as large
as those applied for. In the event of damage caused by a project, the
affected party is entitled to compensation.

On the basis of Finnish legislation, no one directly enjoys the right to
own ground water, but the potential use of ground water found on one's own
land must not be infringed upon. The cost directly arising from the
prohibition against altering or polluting ground water is borne by the
landowner. If the party using ground water wishes to ensure the protection of
ground water through the establishment of a ground-water protection zone, in
addition to regulations prohibiting the altering or polluting of ground water,
that party becomes liable for damage or injury caused to the landowners
affected by this action.

According to Finnish law, the water administration is vested in the
National Board of Waters and subordinate district administration. The country
has been zoned into 13 water districts for regional administration. Three
water courts act as the sole authority granting permits and licences in
water-use cases. Questions involving the protection and use of ground water
are handled within the water administration. In the course of water court
proceedings concerning ground-water use, applicants are required to order
hydrogeological studies and monitoring at their own expense. In addition to
case-by-case supervisory action, the water administration authorities conduct
nation-wide general ground-water research at specially established
ground-water research stations. This research charts ground-water conditions
in general and examines the impacts of action endangering ground water.

France

In France, the exploitation of underground resources is subject to the
granting of a concession by the public authorities if the resources in
question are considered to be of great importance to the country's economy or
independence. Water resources do not fall into that category and, for that
reason, the general rule is that ground water belongs to the owner of the
land, who may, in theory, exploit it as he sees fit. This rule, which is

based on custom rather than on any specific legislation, explains why it was not until 1935 that the first legal text restricting the discretionary use of ground water was adopted, whereas the first measures regulating such use of surface waters date back more than a century. Since 1935 the following additional measures have been adopted:

On the one hand, regulatory measures aimed at controlling the exploitation of ground water and protecting its quality;

On the other hand, within the framework of a general Water Act, a set of provisions to serve as guidelines for ground-water management in the light of the local importance of the water's use.

Under the Civil Code, ground water, like other underground resources, belongs to the owner of the overlying land. The Mining Code restricts this general rule by excluding from this right of ownership those substances requiring a concession (metals, coal and hydrocarbons, certain minerals, and geothermic water), whose exploitation is assigned and controlled by the State.

France does not have a complete set of legal rules specifically applicable to ground water. Applicable provisions, to be found in texts of diverse origin, can be classified under three headings, as follows:

(a) Prevention of overdrawing: The Decree of 8 August 1935 is the first piece of legislation restricting the use of ground water "taking account of the public interest which attaches to the preservation and rational utilization of this resource". The Decree provides that all works executed at more than a certain depth within an area delimited solely by natural boundaries constituting the area as a geological entity shall be subject to previous authorization. This stipulation, at first applied only to the Paris area, where an alarming fall in the aquifer level had been recorded, was gradually extended to several regions where ground water is widely used, particularly for the supply of drinking water. The concern with good management was extended to the whole country by the enactment of the 1964 Act "on water management, water distribution and the control of pollution". Article 40 of this Act specifies that any installation, at whatever depth, having a capacity exceeding 8 m^3/h and not intended to meet the user's own requirements must be brought to the knowledge of (declared to) the administrative services responsible for the management of water resources in order that those services may supervise the installation's operation.

(b) Protection of the user: Article 113 of the Rural Code recalls that any diversion of ground water (and of surface water) for the purpose of meeting collective requirements shall be subject to a public inquiry procedure to enable the installation to be regulated while safeguarding the resource and thus protecting the general interests of the community.

(c) Protection of the quality of the water: This is accomplished essentially by applying the provision of the Public Health Code which stipulates the establishment, around draw-off points intended for the supply of drinking water, of a number of protection areas, described as "perimeters", where constraints are established in relation to the catchment's hydrogeological context (depth of the installation, protection of the aquifer against pollutants from the surface, conditions of discharge into the aquifer, etc.) and of the catchment's physical environment (urbanization, industrialization, spreading operations, dumping and landfill, etc.).

The various protection perimeters are defined by the administration on the basis of a report by an "approved hydrogeologist" (i.e. approved by the Ministry of Public Health). The approved hydrogeologist determines the scope of each protection perimeter (immediate, proximate or extended) and the restrictions on the use of the soil and subsoil to be applied within these perimeters to ensure the protection of the quality of the water collected. The application of such restrictions is subject to a regulatory procedure, including a public inquiry, to bring out the various interests involved at the level of each protective perimeter.

Ground-water protection is also provided for by the Act of 1964 (Decree No.73-218) which regulates discharges capable of impairing water quality by subjecting them to administrative authorization. In addition, there are numerous regulations applicable to various activities (spreading operations, industrial and radioactive wastes, camping, etc.) on a case-by-case basis which are supposed, in principle, to preclude the infiltration of pollutants into aquifers. All in all, it is not always easy to make rational use of all these provisions, which are applied by different administrative departments, often with competing development objectives.

The 1964 Act, that has already been mentioned, endeavours nevertheless to provide a general framework for policies which may vary from region to region depending upon the importance and vulnerability of the ground-water resources. As already stated, by extending the declaration requirement to the whole country, the Act has made it possible to obtain a reasonably clear overall picture of the draw-offs although the situation with regard to agricultural draw-offs for irrigation purposes still gives rise to some difficulty. With regard to quality, the Act stipulates that any action (discharge, dumping, disposal, etc.) likely to impair the quality of the ground water may be regulated or prohibited, the scope of any restrictions imposed being left to the discretion of the local authorities in the light of the importance of the water's intended uses and of its quality.

In addition to these regulations, which are intended to discourage abuse, efforts have been made to make the users and administrative departments concerned share responsibility for the management of water resources by setting up catchment area bodies, namely, a river basin committee comprising elected representatives, users and members of the administration, to be responsible for defining a policy for the development of these resources and deciding upon the operations to be undertaken, and a river basin financial agency, on which the administrative departments responsible for the policing and management of water resources are represented. The policy worked out at the level of the country's six major river basins is financed partly out of charges levied upon all users - municipalities, industries and farmers - in proportion to the amount of water they use (consumption charge) and to the impairment of water quality caused by each user (pollution charge).

The policies of the River Basin Committees with regard to ground water tend to be stricter in the areas of greatest use. In the north, east and south-west of France and in the Paris region, extremely detailed arrangements concerning the level of aquifers and changes in their quality have been instituted, including, in particular, steps to limit ground-water use in certain areas to certain specific purposes and to place those areas under the sole responsibility of the public authorities (State or local government). In these regions, the use of resources forms part - at least locally, as in the Paris region - of medium-term and long-term development plans which also cover surface waters. In the south of France, ground-water use is regulated in a more traditional and localized way and the major development schemes relate essentially to surface waters.

German Democratic Republic

Important legal documents adopted by the Government of the German Democratic Republic, including the new Water Law of 2 July 1982, are aimed at the rational use of water in general and the protection of drinking water resources in particular. The necessary protection of ground water as the major source of drinking water supply must be ensured by preventive and curative control measures. This applies especially to contaminants such as nitrate and other nitrogen compounds, persistent organic substances, heavy metals and inorganic salts.

The Land Improvement Law adopted by the people's chamber of the German Democratic Republic in 1970 had stipulated that: "Handling of substances that may cause water pollution has to be organized in such a way as to exclude health injuries of citizens and economic damages as well as to avoid negative effects on water resources, flora and fauna." This general statement has been embodied in the new Water Law of 1982 as follows: "Water protection is a social task of State authorities, factories and citizens."

The basis of prophylactic ground-water protection is the Water Law of 1982 where principles for protection zones and reservation areas are formulated with general prohibitions and limitations to be followed in using such areas for municipal, industrial and agricultural purposes. (It should be noted that about 13 per cent of the German Democratic Republic territory and about 14 per cent of arable land is covered by protection zones. From these figures the particular effect of protection area policies on agricultural production becomes obvious.)

The principles of protection zones are formulated more in detail in the German Democratic Republic Standard TGL 24348 "Use and Protection of Water, Drinking Water Protection Areas". This Standard is divided into four parts as follows:

Part 1: General Principles,

Part 2: Water Protection Areas for Ground Water,

Part 3: Water Protection Areas for Surface Water,

Part 4: Marks in the Country, Signs in Maps.

Part 5 on drinking water reservation areas is under preparation.

In addition to the above-mentioned Standard, Standard 24345 "Use and Protection of Water, Principles of Handling Organic and Inorganic Fertilizer" must be mentioned. The Standard is particularly aimed at protection of ground water against the application of agricultural fertilizers in areas outside of protection zones. Similar regulations exist also for the application of biocides as well as for agricultural irrigation with clear and waste water (land treatment of waste water). For other contaminants and accidental spillage, special protection and sanitation programmes will be elaborated. Research work has been initiated on the application of so-called active substances for controlling biochemical processes in soil and ground water.

Federal Republic of Germany

Legal provisions regarding use and protection of ground-water resources are contained in a number of laws and regulations, particularly in the Act on the Regulation of Matters relating to Water (Federal Water Act). Relevant articles of this Federal Water Act are quoted below.

In the introduction, article 1 stipulates, inter alia, that:

"(1) This Act shall apply to the following waters

    1.   water which permanently or temporarily stands or flows in beds or which flows from natural springs (surface waters),

    1a.   the sea between the coastline at medium flood water level or the seaward limitation of the surface waters and the seaward limitation of the coastal sea (coastal waters),

    2.   underground water.

(2) The Laender may exclude from the provisions of this Act small waters which are economically of minor importance as well as springs which have been declared to be healing mineral springs ..."

In Part I on the general provisions relating to waters, the basic principle stipulates, inter alia, that:

"... (3) Ownership of land does not entitle [anyone]

    1.   to water utilization which requires a permit or a licence according to this Act or to the Water Laws of the Land, ..." (art. 1a) and

"(1) Waters may not be used without a permit (art. 7) or a licence (art. 8) issued by the authorities ..." (art. 2)

For the purpose of the Act, ground-water uses are defined in article 3 as:

"... 5.   the discharge of matter into underground water,

    6.   the withdrawal, bringing or drawing-off to the surface, or the diversion of underground water.

(2) The following shall also be deemed to constitute uses of water

1. the impounding, draw-down or diversion of underground water by
    means of installations designed for, or suitable for, such
    purposes,

2. any measures which are likely to cause, either permanently or
    to a not merely insignificant degree, harmful changes in the
    physical, chemical or biological constitution of water ..."

Articles 4 to 12 contain detailed provisions regarding permits or
licences in particular on conditions of use, reservation, refusal,
requirements for discharging waste water, licensing procedure, concession for
early start, subsequent decisions, exclusion of claims and limitation and
withdrawal of licence. Articles 13 and 14 concern uses of water by
associations and approved plans and mining operation plans, respectively.
Articles 15 to 17 refer to old rights, old authorities and other old uses.

Permit-free uses are covered in article 17 (a) and include, <u>inter alia</u>:

"... (a) the temporary removal of water from a water [source] and the
          discharge of the water back into a water [body] by means of
          movable installations, as well as

(b) the temporary introduction of substances into a water [body],

     if this does not prejudice others at all or only slightly, if
     no injurious change to the properties of the water is to be
     expected and if no other impairment to the water situation is
     to be expected. The project is to be reported to the competent
     Water Authorities ..."

Articles 18, 18 (a) and 18 (b) provide for settlements in respect of
rights and authorities, obligation to and plans for disposing of waste water
as well as for erection and operation of waste-water installations.

Article 19 provides for establishing water protection areas:

"(1) So far as may be necessary in the interests of the common weal

1. that certain waters should be protected against detrimental
    effects in the interests of the now existing or of any future
    public water supply, or

2. that underground aquifers should be recharged, or

3. that the harmful effects caused by the run-off of rain water
    should be prevented,

water protection areas may be established.

(2) In these water protection areas

1. certain activities may be forbidden or permitted only to a
    limited extent, and

2.   the owners and persons enjoying the beneficial use of real property may be obliged to accept certain measures, including measures taken for the observation of the water and the soil.

(3) If an order made under paragraph 2 amounts to expropriation, compensation shall be paid for this; ...

(4) The establishing of a water protection area shall be done by way of formal procedure."

Articles 19 (a) to 19 (l) deal with pipeline systems which are destined for the conveyance of materials deleteriously affecting water and with facilities for storing, filling and handling substances constituting a water hazard particularly regarding impositions and conditions, refusal, limitation and withdrawal of licences, etc.:

"(1) For the protection of waters, in particular for the protection of ground water, the licence may be granted subject to the imposition of conditions ... The licence can be made subject to a time-limit. Even after the issue of a licence, impositions with regard to quality requirements and to the operation of a system shall be permissible in cases where it is to be feared that contamination of the waters or any other disadvantageous alteration of their properties will occur.

(2) The licence shall be refused if, by the establishment or by the operation of the pipeline system, contamination of the waters or any other disadvantageous alteration of their properties is to be apprehended and can neither be prevented nor compensated by impositions. In the case of pipeline systems which cross the frontiers of the Federal Republic of Germany, the licence can also be refused if the concern is caused by parts of the system which are erected or operated outside the area of applicability of this Act." (art.19 (b)).

Moreover,

"(1) Facilities for storing and filling substances constituting a water hazard must be designed, installed, erected, maintained and operated in such a manner that a contamination of the waters or any other disadvantageous alteration of their properties is not to be feared.

(2) Facilities for transloading substances constituting a water hazard must be of such a quality and must be installed, erected, maintained and operated in such a manner that the best possible protection of the waters against contamination or other disadvantageous alteration of their properties is obtained.

...

(4) Legally valid regulations of the Laender covering the storing of water-hazardous substances in water protection areas, waterhead protection areas, inundation or project areas remain unaffected ..."
(art.19 (g)).

In addition,

"The operator of facilities according to article 19 (g), paragraphs 1 and 2 has the obligation to check for leaks continually and to supervise constantly the operating status of the safety devices. In individual cases, the competent authority may require that the operator conclude a contract of supervision with a company which has been approved according to the laws of the Land, if the operator himself does not possess the necessary expert knowledge or does not have his own specialized personnel. Moreover, according to the laws of the Land, he shall have installations checked for proper conditions by officially recognized experts, in the following cases:

1.   before starting up the plant for the first time or after a major change,

2.   at the latest every five years, and in the case of underground location in water- and waterhead-protection areas at the latest two and one half years after the last examination,

3.   before starting up the plant again after it has been standing idle for more than one year,

4.   if a check is ordered because it is feared that a danger to water exists." (art.19 (i))

Compensation for any damage arising to property is provided for under article 20.

Regarding supervision:

"(1) Any person who uses any water in excess of the normal usage, or who has submitted an application for the grant of a permit or licence, shall be obliged to accept official supervision of the facilities, devices and procedures which are of importance for the use of the water. For this purpose, and in particular to check if a requested use may be approved, what conditions of utilization and other conditions have to be imposed in this connection, whether the use is within the permissible limits and whether orders have to be subsequently issued on the basis of article 5 or of supplementary regulations according to the laws of the Land, he shall allow at all times:

1.   access to business land and business rooms during working hours,

2.   access to dwelling rooms as well as to business land and business rooms outside working hours so far as the examination is necessary for the public safety and in order to prevent serious danger, and

3.   access to land and installations which do not belong to the directly bordering fenced property according to items 1 and 2, ...

The owners and possessors of land on which the plants are manufactured, erected, installed, set up, maintained or operated shall allow access to the land, grant information and permit technical investigations and tests about pollution of the water or any other disadvantageous alteration of its properties. The same shall apply to the conveyance of liquids and gases by means of pipelines."

Articles 21 (a) to 21 (f) deal with the appointment of company agents for water protection, specifying their duties, the obligation of water users to appoint such agents, their opinion on investment decisions, their right to report, etc. Article 22 is concerned with liability for changes in the constitution of water.

Part II of the Water Act covers regulations for surface waters (articles 23 to 32), while Part III sets out provisions governing coastal waters (articles 32 (a) and (b)). Part IV deals with regulations for underground water. The latter are quoted in the following:

"Article 33

Utilizations not subject to permits

(1) Neither a permit nor a licence is required for the withdrawal, bringing or drawing-off to the surface or the diversion of underground water

    1.   for domestic purposes, for farming purposes, for watering cattle outside the farm or for use in small quantities for some temporary purpose,

    2.   for the purpose of the normal drainage of land used for farming, forestry or gardening purposes.

(2) The Laender may lay down, either as a general rule or for specific areas,

    1.   that a permit or a licence shall be required in the cases referred to in paragraph 1,

    2.   that neither a permit nor a licence shall be necessary for the withdrawal or bringing or drawing-off to the surface or the diversion of underground water in small quantities for trade purposes or in connection with farming, forestry or gardening operations exceeding those specified in paragraph 1.

Article 34

Preservation of the Purity of Water

(1) A permit for the introduction of matter into underground water shall be granted only where there is no danger of that water being harmfully polluted or of any other detrimental change in its qualities.

(2) Materials shall be stored or deposited in such a way that there is no danger of any harmful pollution to the underground water or of any other detrimental change in its qualities. The same shall apply to the conveyance of liquids or gases through pipes.

Article 35

Earth Workings

(1) So far as is necessary for the regulation of the water supply, the Laender shall lay down that any works which penetrate into the ground below a certain depth shall be subject to supervision.

(2) If underground water is opened up unintentionally or without authority, an order may be made for the opening up to be made good, if considerations of water supply make this necessary."

Part V of the Water Act covers general water planning and water registers. Part VI deals with penalties and fines and Part VII sets out final provisions.

The general provisions of the Federal Water Act relating to piping systems for the transport of liquids presenting a hazard to waters are completed and specified in more concrete terms in the water laws of the Laender. For example, a guideline on pipelines for the transport of hazardous substances has been introduced for the equipment, construction and operation of piping systems subject to licensing under water laws.

The relevant ordinances passed by the Laender regarding equipment and installations for storage, loading and handling of substances presenting a hazard to waters (1981) and on the approval of special enterprises concentrating on equipment and installations for storage, loading and handling of substances hazardous to waters ("Fachbetriebsverordnung", 1982) specify these demands in more detail and give further requirements. Through these two ordinances, the general safety requirements of the Federal Water Act are translated into four concrete steps: (a) Design, layout, maintenance and operation of equipment and installations for storage and loading of substances presenting a hazard to waters must be such that any pollution of waters need not be feared. The best possible protection is required for equipment and installations used for handling substances that present a hazard to waters. (b) A preliminary assessment by the responsible authorities is required for such equipment and installations (technical aptitude testing or type testing). This preliminary assessment is not necessary for simple or conventional equipment or for installations on which the assessment is based as specified under water law. (c) The equipment and installations used must be inspected at intervals matched to the risk potential. (d) Enterprises which, on a commercial basis, install, erect, repair, maintain and clean the above installations are subject to approval.

These provisions do not apply to production plants. This sector is governed by the Federal Immission Control Law. This law stipulates that installations which by their nature or operation are particularly liable to cause harmful effects to the environment or otherwise endanger or cause considerable disadvantage or nuisances to the general public or to the vicinity must be subject to licensing. Such installations must be established and operated in such a way that harmful effects cannot be caused to the environment nor pose other dangers, considerable disadvantages or considerable nuisances to the general public and the vicinity. The term "other dangers" also includes any contamination of ground water that may have detrimental effects on human health or which may damage considerably material goods.

Ground-water quality protection is also ensured by standard procedures for instituting and dimensioning protection zones. They have been published together with scientific and engineering background materials. The technical instruction manual "Richtlinien für Trinkwasserschutzgebiete (Grundwasser)" differentiates within one water protection area up to four separate protection zones at increasing distances from a well. Bans and restrictions in each protection zone can be found in the manual cited as well as in scientific literature on the subject. The system provides that all restrictions in a zone of lower protection quality be valid also in more protected zones. Various activities are covered, from agriculture through construction and industry to holiday resorts: some 40 or 50 categories of activities.

Besides the water laws already mentioned, some product-related regulations serve to protect ground waters. The "Rules of Handling Water - Endangering Substances" contain groups of potentially dangerous products and compounds as well as regulations aimed at water protection. Pipe systems for such substances require a special approval procedure. Likewise installations for their storage or handling must fulfil special requirements, and companies or persons need official permission for such activities.

The "Plant Protection Act" and related regulations contain rules concerning the application of pesticides in water protection areas. There is a regularly updated list of pesticides with a classifying system of bans and restrictions based on the different water protection zones and the respective properties of the substances.

Regarding control of ground-water pollution arising from agricultural activities, the Federal Republic of Germany recently enacted the following legal regulations:

(a) Regulation on sewage sludge - This regulation lays down some basic principles which require that sewage sludge be used in agriculture in a manner which does not adversely affect the environment; these basic requirements particularly limit the pollution load to seven heavy metals both in the soil and in sewage sludge. The aim is to secure the agricultural use of sewage sludge on a long-term basis without damaging soil and waters.

(b) "Regulation on Semi-liquid Manure" (Circular Decree issued by the Lower Saxon Minister of Food, Agriculture and Forestry) - Attempts at setting up the Regulation on Semi-liquid Manure for the entire area of the Federal Republic of Germany had failed in 1979. The Federal Laender are now in a position since the re-enactment of the Waste Disposal Act to make their own legal provisions in order to control nitrate concentrations in certain areas caused by an excessive use of semi-liquid manure. Lower Saxony was the first Federal Land which recently adopted a so-called "Liquid Manure Regulation". According to this Regulation it is permitted to apply a maximum of three units of dung per hectare of cultivated land and per year (with the exception of the time between the end of October and February), provided that the "usual extent of agricultural fertilization" is not exceeded.

Greece

Greek legislation covering water management is based on Decree 608/1948 where in article 1 a distinction is made between public and private water. Private water is considered to be the water which springs out on privately owned land as well as that located under private ground, the latter refers to underground water which is brought up to the surface by the owner of the land using pumps or other technical means. It comprises also public waters whose use has been allocated to an individual for private purposes, the rain water which collects within the borders of the privately owned land, and finally the water which flows in pipes, whether for private or municipal purposes, as long as the pipe runs through private land. Public waters are considered to be all the flowing or State waters of the country, whether surface or underground, except those defined above as private.

Article 2 of the same decree specifies that any right on public water assigned for private use be restricted to that use only, it is forbidden to allocate surplus water, which is considered public, to any other person.

As regards irrigation water, article 8 of the same Decree 608/1948 specifies that the Minister of Agriculture may regulate its use and management according to the type of cultivation and to the extent, duration and repetition of the irrigation cycles. The Minister could also forbid the digging of new irrigation wells in certain areas, such as for instance in the greater Athens area, if a given area is considered to be over-exploited by existing irrigation wells.

Article 9 provides for possible compulsory expropriation of any surplus private water which is not needed by the owner of the land. In fact, any private water could be taken over by the State in cases of emergency and its use allocated to agriculture in the most beneficial way (article 10).

Another reference to water management is made in Decree 439/1945. Under article 1 of this decree, restrictions are imposed regarding excavation of irrigation wells and pumping of underground water in order to secure regular water supply for cities, towns or villages. The outline of such a restriction zone, the permissible quantities of water to be pumped from wells already operating and the duration of these restrictions are decided upon jointly by the Ministers of Agriculture and of Public Works.

Decree 560/1968 deals with emergency provisions for supplying water to residential areas. In such cases, the municipal authorities can take over all the surplus private waters (as defined above) and distribute them according to the specific needs of the population.

Finally, article 202 of Decree 1065/1980 permits the expropriation of any private property, if this will facilitate works for water supply or for collection, transportation, distribution and improvement of water.

The above review of some aspects of Greek legislation regarding water management demonstrates the concern of the State about this sensitive issue, it may also show how much more needs to be done. More specifically, modern legal arrangements which would cover globally all cases of water management and protection need to be introduced.

For this purpose, the Ministry of Energy and Natural Resources is preparing the new "Law for Water" which defines water as one of the main elements of life which has to be protected. According to this law, all the institutions and organizations authorized to define and control the use of water must co-operate, in order to minimize the possibilities of overlap. Within the framework of this law, all existing relevant legislation will be revised. Another objective of the same law is to establish the decentralization of the water management authorities. Furthermore, citizens are expected to become more environment-conscious when they realize that the protection of ecological systems is the responsibility of all.

The Ministry of Town Planning and Environment is now preparing a new law, which is expected to bring about radical changes in the existing policies of environmental protection. The new philosophy is to prevent pollution rather than try to reduce it. Simultaneously, an effort has been made to provide legal documentation covering the entire issue of environmental pollution, since the existing legislation is fragmentary and often insufficient. It certainly does not provide for heavy fines to be imposed. This new law defines environmental protection as the series of measures and activities which generally aim at the preservation of the environmental balance while improving the quality of life of both present and future generations and achieving a balanced development of society. Particular emphasis is placed on the need to protect all the natural resources of the country; special reference is, of course, made to surface and underground waters as well as to the seas.

Hungary

Legislation on the protection of subsurface water resources had been initiated in Hungary around the second half of the nineteenth century (Act XIV of 1876). The importance of the delineation of protective areas around wells, mineral springs and spas was emphasized by Act XIII of 1885, which introduced overall legislative order in the field of water law in the country. This act - one of the oldest in Europe - protected the quality of subsurface waters there for about 70 years. Its modernization was initiated in 1961.

During the 1960s a rigid policy was adopted to define protective areas. Dimension, form and other characteristics of such areas were determined by water authorities and then passed on to the planner. The scheme did not distinguish between small and large, well-, less-protected or unprotected areas, and water works. The goal of the legislation was the protection of the water-producing project itself and its immediate surroundings. This practice was initiated at a time when organic manure was the main source of pollution and protection was thus limited to this material. In the past 20 or 25 years the factors endangering the quality of subterranean waters have increased in a spectacular way. As a consequence, damages caused by these factors became more frequent than earlier.

It became evident, however, that rigid legislation would not promote qualitative protection. Experience shows that, for a reliable delineation of the protective area of a water resource project, it is indispensable to have a thorough knowledge of the geological, hydrogeological, hydraulic, etc. characteristics of the given site. Practical solutions to this problem can be worked out when the overall situation for geological protection has been studied. Subsequently, natural conditions, effective technical solutions and authoritative rulings can be successfully combined.

The present legal system in Hungary foresees that management of ground-water resources forms an integral part of the general water legislation, with its corresponding rulings referring equally to matters both of surface and subsurface waters. While some ground-water features allow certain specific uses (thermal, mineral waters, etc.) they may occasionally fall also under the jurisdiction of specific regulations, the elements of the latter, however, will always reflect the principles laid down in general water legislation.

Water legislation in Hungary does not recognize any priority to any user of the ground-water resources but grants equal rights to all users of these waters. In view of this, the legal principle is that the available resources will be allocated by the State through its executive power by granting water rights for the use of certain well-defined parts of these waters for the given purposes, irrespective of the subjective character of the users.

According to Hungarian water legislation, any use of ground-water resources is subject to authorization. This implies all water production works and water uses, irrespective of whether or not these activities serve the user(s) own purposes or are done on some commercial basis. (No licence is needed, however, for using the ground water originating from the uppermost aquifer if it is for domestic purposes.) Relevant legislation leaves only a narrow sector open for free practices that do not endanger the general interest of society.

Italy

The Italian Parliament approved and promulgated a law (No.319 of 10 May 1976) intended to protect national water resources against pollution. The law concerns all types of water use, emphasizing the technical and legal aspects of pollution control. The financial aspects are specified in detail, in order to obtain a fair allocation of treatment charges. The application of the law has involved detailed research into basic data pertaining to knowledge of water pollution and means to re-establish acceptable qualitative levels, within the wider context of environmental protection and of rational use of natural resources.

The law has the following purposes:

(a) Control of discharges of waste water whether public and private, direct and indirect, into surface waters and ground waters, both fresh and saline, public and private, into sewers, onto the ground and into the subsoil.

(b) The formulation of general criteria for the use and the discharge of waters.

(c) The organization of public utilities: water supply system; sewer and waste water treatment plants.

(d) The drawing up of a general plan for water quality improvement on a regional basis.

(e) The inventory of water resources in terms of quality and quantity.

For the application of Law 319, the proper jurisdiction is specified of central Government, region and provinces. The central Government has, first of all, to promote consulting and general co-ordination and to point out criteria and methodologies for carrying out the survey of "water bodies". A general master plan for water-quality improvement should also be promoted, taking into account co-ordination with regional planning.

Initially it was necessary to provide a general definition of "water body" as "any mass of water which, regardless of its size, has its own hydrological, physical, chemical and biological properties and which is or could be used for one or more purposes".

All permanent and temporary deposits of water in the subsoil, being of sufficient quantity for at least seasonal use, are known as ground waters. This category includes phreatic water-bearing strata, as well as deep strata whether under pressure or not, contained within porous or fractured soils and, to a lesser degree, still masses of water entrapped within deep rocky formations. It also includes springs, both concentrated and diffused, which may be subaqueous, as such springs indicate the presence of ground water.

In compliance with Law 319, underground water deposits and springs particularly vulnerable to pollution, direct or indirect, must be studied first. With regard to ground water, the delimitation of each hydrogeological basin is particularly important: its boundaries are related to the geological structure of the rocky formations. Consequently, strata with water-bearing aquifers are to be studied in a comprehensive manner, together with their recharge and discharge areas. As for springs, the description of the water body must not be limited to the geographical and topographical aspects of the location of the spring, but include also its parent water body in the subsoil.

Investigation of the hydrological characteristics is closely related to the geological surveying necessary to determine the location of underground water-bearing reservoirs and hydrogeological basins. Such surveys must include the entire basin, in order to determine the pattern and condition of the water-flow, the relation to surface hydrography and the overall hydrological balance. With regard to the physical, chemical and biological properties of underground waters, in general a limited list of basic analytical data must be compiled covering: temperature, hardness, conductivity, basic ionic types (Na, K, Mg, Ca, Fe, Cl, SO_4, NO_3), and bacteriological indexes (fecal coli). In special instances, other data may be obtained pertaining to specific aspects of pollution.

Further aspects of the Law relate to the regulation of the dispersion of liquid wastes in the soil and into the subsoil. It is provided that, during the last phase of treatment, liquid waste may be dispersed either into the soil and into the superficial strata, or into deep and isolated geological formations. The use of intermediate strata, which may become unforeseeably connected to usable water-bearing strata, is prohibited. Only processed waste-water, which may be entirely purified by exposure to natural phenomena occurring in the atmosphere, may be discharged into the soil.

Substances containing waste materials which are pollutants, particularly difficult to process and which must be isolated from the biosphere, may be discharged into deep geological formations.

With regard to dispersion into the soil or superficial strata, a distinction must be made in the case of soil used for agricultural purposes. This category includes any area producing crops which can be used, directly or indirectly for animal or human consumption, or for industrial processing, or which is the object of commercial activity. Although the protection of water-bearing strata is of primary importance, in the case of discharges into agricultural soil, special conditions are set in order to prevent damage to crops and consequently to the health of both animals and human beings.

The main uses must be listed for each underground water body.

As final disposal for sludge, the Law provides for the use of the following sites, in accordance with specific characteristics and restrictions:

- Soil used for agriculture

- Soil not used for agriculture

- Subsoil

- Territorial seas

- Extra-territorial seas.

For each of the above, the terms and means of disposal have been determined.

The application of the Law provides a survey of the disposal of sludge, to be carried out by the regions, in order to ascertain that the treatment, means and sites of disposal conform to all specifications.

The Law is particularly concerned with the means by which water is withdrawn from springs, underground water-bearing strata (by wells or filtering channels) and surface waters (such as lakes and rivers), with or without benefit of reservoirs.

All technical and fluodynamic aspects pertaining to the various means of withdrawal must be taken into account when planning for new withdrawal systems and for the extension of the already existing ones. In order to avoid polluting water supplies and jeopardizing their further use, specific investigations must be carried out individually for each case. If necessary studies must be made by specialized research laboratories and institutes, according to the most advanced techniques for information analysis.

Luxembourg

The Ground-water Protection Act of 9 January 1961 provides as follows:

"Article 1. Any new ground-water catchment and the installations pertaining thereto shall be subject to prior authorization by the Minister of the Interior.

Work in progress at the time of the entry into force of this Act shall likewise be subject to such authorization.

The term 'ground-water catchment' shall include all devices to tap springs at the point of emergence, wells, drillholes, boreholes, culverts, drains and, in general, all works and installations whose purpose or result is to effect a ground-water extraction."

"Article 2. The conditions for establishing a general inventory of the country's ground-water resources shall be laid down in regulations issued by the public administration."

"Article 3. The following shall be dispensed from the authorization referred to in article 1:

    (a) All ground-water draw-offs of a depth not greater than 20 metres effected in a non-artesian aquifer and operated manually.

    (b) Drainage installations and aquifer drawdown operations which do not cause the water table to fall more than 2 metres below the natural ground level."

"Article 4. The following shall be treated in the same way as new ground-water draw-offs:

    (a) Any alteration as a result of which an existing water draw-off would cease to meet the conditions for exemption listed in article 3.

    (b) Extensions or modifications to any ground-water draw-off not covered by article 3.

    (c) The bringing into use of old ground-water catchments which had remained without regular use for a continuous period of five years."

"Article 5. The development and commercial exploitation of quarries in Luxembourg sandstone are subject to prior authorization by the Ministers of the Interior, Public Health and Public Works."

"Article 6. The landowner, works contractor and operator shall be required to apply for the authorizations referred to in articles 1 and 5."

"Article 7. Authorizations may be withdrawn or suspended if the conditions subject to which they were issued are not complied with; the same applies if any new conditions which the competent Ministers may impose at any time are not complied with."

- The Act pertains to all ground-water abstractions, not just those of water supply companies. The Act on Ground-water Abstraction for Public Water Supply will be withdrawn as soon as the new Act takes effect. Nevertheless, licences granted under the old law may be continued if not withdrawn by the provincial authorities.

- The new Act provides not only for licensing of ground-water abstraction but also licensing of ground-water recharging with surface water. Furthermore, the Act instructs the provincial administrations to prepare ground-water management plans.

Planning has of course always been part of management, but legal frameworks were lacking. Now planning forms a part of virtually every law on water management or land use.

The features in common with the existing law are as follows:

- Although ground-water abstractions for which permissions are granted by the provincial authorities are not, as a matter of course, in the public interest, owners of real estate in the vicinity of the abstraction are obliged to be indulgent, while they are entitled to indemnification if the abstraction is prejudicial to their interests.

- If the value of their real estate decreases too much, they may even claim for the property to be taken over by the licensee. As under the existing law, a commission of experts on the subject may also be asked to assess the damage caused by ground-water abstractions.

It is feared that prescriptions set out in too many laws will discourage industrial activities. A certain stimulus to licensing authorities is provided by the Environmental Protection Act (general provisions). By virtue of this Act, authorities dealing with applications for licences are obliged to state their decisions within a fixed time. Most other acts concerning environmental protection are covered by the Environmental Protection Act. Whether the new Ground-water Abstraction Act will be covered by these general rules is as yet uncertain. This also explains why the Ground-water Abstraction Act, issued in 1981, is now still only partly in force. Another problem has arisen since the Government started to remove regulatory obstacles from the path of commerce by installing a commission for "deregulation", with the intention of withdrawing legal regulations by which the yields of commercial enterprises could be diminished.

Switzerland

The first Federal Act on water protection dates from 16 March 1955. It very soon proved inadequate.

A new Federal Act on the Protection of Water against Pollution (abbreviated as APWP) more detailed and stricter was promulgated on 8 October 1971 and, with some slight amendments, is still in force today. Its basic object is to avoid any change in the physical, chemical and biological properties of exploitable ground water, irrespective of its present use, and to prohibit the disposal of pollutants by allowing them to infiltrate into the subsoil.

APWP introduces three subdivisions of Swiss territory:

<u>Water protection sectors</u>, or A, B, and C areas. In these areas, general protection is given to ground-water resources on the basis of their greater or lesser value from the point of view of water supply and of their vulnerability;

<u>Ground-water protection areas</u>, or S areas, established around water catchments used for the supply of public drinking water;

<u>Ground-water protection perimeters</u> (also S areas), where new catchments may be established and appropriately protected in the future.

The stringency of the conditions imposed on land use in the interests of ground-water preservation increases in the following order: C-B-A-S areas.

The second chapter of APWP contains detailed provisions for the prevention of pollution, as follows:

"Article 13. All persons shall be in duty bound to endeavour to prevent any form of pollution of surface and ground waters with such diligence as the circumstances may require.

Article 14. (1) The direct or indirect introduction or dumping in water of any solid, liquid or gaseous substance capable of polluting it is prohibited. It is likewise prohibited to dump away from water any substance which might pollute it.

(2) It is forbidden to eliminate pollutants by permitting them to sink into the subsoil. The competent cantonal authority may authorize exceptions where there is no risk whatsoever of pollution of surface or ground water ..."

"Article 15. Liquid or gaseous substances, particularly sewage, coming from the drainage systems of localities, dwellings, building sites, industrial or artisanal enterprises, farms, ships or elsewhere, may not be discharged into water unless they have been treated in accordance with cantonal regulations. The discharge of sewage must be authorized by the competent cantonal authority.

Article 16. (1) It shall be the responsibility of the cantons to ensure that all methods of elimination by discharge or by infiltration which may cause pollution are adapted to the requirements of water protection or discontinued within a period of 10 years from the entry into force of this Act. The cantons shall prescribe time-limits having regard for the degree of urgency in each case and in conformity with the cantonal water sanitation plan. Longer time-limits may, exceptionally, be granted in the case of insignificant discharges or infiltrations.

(2) The Confederation shall approve the cantonal water sanitation plans and supervise their execution; in particular, it shall take whatever steps are needed to ensure that the prescribed time-limits are observed.

(3) Landowners who discharge untreated or insufficiently treated liquid residues directly into water or allow them to sink into the soil must report the fact to the competent authority within a period of one year from the entry into force of the present Act, specifying the nature and quantity of the wastes thus eliminated. Landowners in possession of a valid authorization from the canton are dispensed from this obligation."

The articles which follow concern the principles governing sewage water, the evacuation and treatment of sewage, permits to build inside and outside the perimeter of the master plan for the sewers, special methods of eliminating sewage, and the examination and preservation of water purity.

"Article 22. (1) The Federal Council shall draw up regulations concerning:

(a) The periodic examination of surface water and ground water;

(b) The nature of the waste water discharged into the waters;

(c) Regular inspection of public and private installations for the treatment of sewage water;

(d) The characteristics of residues from the waste-water plants, their use or their adequate disposal.

(2) After hearing the views of the cantons, it may enact ordinances containing special rules for preserving the purity of certain waters."

Article 23 relates to products, substances and production processes having harmful effects:

"Article 23. (1) The Federal Council shall draw up provisions concerning:

(a) Products which, by reason of their mode of use, penetrate into waters and may, because of their composition, have harmful effects upon the operation of installations for the disposal and treatment of sewage waters or damage the quality of water;

(b) The elimination or conversion of substances which may damage the quality of water;

(c) Production processes giving rise to waste water containing non-degradable toxic substances;

(d) Products which, by reason of their mode of use, are converted into wastes or refuse whose satisfactory elimination in accordance with this Act is not possible or is disproportionately costly.

(2) The Federal Council may, where necessary, prohibit the manufacture, use or importation of products or the application of production processes mentioned in paragraph (1) of this article."

Articles 24 to 26 contain provisions concerning substances capable of impairing the quality of water, reservoirs and decanting points, and the overhaul of storage installations, as follows:

"Article 24. (1) The manufacture, treatment, decanting, transport and storage of substances capable of impairing the quality of water, particularly solid or liquid fuels, shall be contingent upon the availability of safety structures or installations facilitating water protection; such structures and installations shall be regularly inspected. ..."

"Article 25. (1) The Federal Council shall draw up regulations concerning the selection of locations which may be authorized for installations serving for the storage, transport, decanting or treatment of liquids capable of impairing the quality of water, the layout and equipment of such installations, and the inspection of the materials employed.

(2) An authorization by the competent cantonal authority shall be required in order to construct, convert or enlarge storage installations or installations used for decanting or treating substances capable of impairing the quality of water. ..."

"Article 26. (1) Any work involved in the overhaul of storage installations may be carried out only by enterprises in possession of a federal authorization delivered by the canton of the enterprise's legal domicile or headquarters. Such authorization shall be granted to enterprises employing qualified personnel and having access to the necessary equipment. The Federal Council shall issue regulations to govern the execution of the work. ..."

Article 27 provides the following:

"An authorization by the canton shall be required for depositing solid matter in waters or in the vicinity.

(2) The cantons shall ensure that solid wastes originating from households, artisanal enterprises or industry are collected and disposed of by dumping in specially prepared tips, by composting, by incineration or in any other manner, provided that no danger of pollution results therefrom. ..."

Articles 29 to 32 relate directly to ground water and are reproduced below in full:

"Article 29. (1) The cantons shall take the steps needed to protect exploitable ground-water aquifers.

(2) They shall subdivide the canton's territory into water-protection sectors, in accordance with Federal Directives and with due regard for the hazards involved.

(3) The construction or conversion of installations and the execution of work, especially diggings, in sectors considered by the cantons to be particularly threatened, shall be contingent upon an authorization by the competent cantonal authority which, in each case, shall stipulate the protective measures to be taken or impose bans.

Article 30. (1) The cantons shall ensure that protection areas are established, as necessary, around ground-water catchments.

(2) The owners of ground-water catchments shall be responsible for collecting data to facilitate the rational delimitation of protection areas, the acquisition of the requisite real-estate rights and, possibly, the payment of compensation for restrictions imposed upon land use. For the acquisition of real-estate rights, the cantonal government may grant to the catchment owners the expropriation right under article 9 of this Act.

Article 31. (1) The cantons shall delimit the perimeters which play an important role in the future use and future artificial enrichment of ground-water horizons. Within these perimeters, installations shall not be established which might pollute the ground water or cause damage to future installations for its use or enrichment.

(2) Where compensation has to be paid, it may be charged to the future owners of ground-water catchments and enrichment installations.

Article 32. (1) Anyone wishing to extract gravel, sand or other material from quarries or in surface waters must be in possession of an authorization by the canton.

(2) In water-bearing strata whose ground-water horizons lend themselves to water-supply purposes as regards both quantity and quality, excavating below the water-level to extract gravel, sand or other materials shall be prohibited. Authorization to extract gravel, sand and other materials above the level of the usable ground-water horizon may be granted provided that a protective layer of materials, the thickness of which shall be prescribed in accordance with local conditions, is maintained above the highest level attainable by the ground-water horizon."

The Act has been supplemented by various implementation orders concerning, for example, the treatment of sewage, the protection of water against liquids which may damage its quality, or the spreading of sewage sludges. The Act has also formed the subject of technical commentaries in the form of directives or recommendations relating to:

The building of roads,

The preparation of dumping tips,

Determination of catchment protection areas,

Protection of water in agricultural areas, and

Utilization of the heat of water and of the soil.

The implementation of Federal water protection laws is the responsibility of the cantons, which refer to those laws in drawing up their own legislation. Thus, cantonal ground-water protection policies throughout Switzerland are organized around the principle of preventing pollution.

Turkey

Ground-water management in Turkey is carried out according to a specific law on ground water of 16 December 1960. According to this law, ground water like other water resources is under the sovereignty and possession of the State. All the activities regarding research on ground water are subject to the provisions of the above-mentioned law.

United Kingdom

In England and Wales use and protection of water resources, including ground water is the responsibility of the 10 water authorities established under the Water Act 1973. Different organizational and statutory arrangements but with similar objectives apply in Scotland, and Northern Ireland under the Water Act (Northern Ireland) 1972.

General water conservation duties of the 10 water authorities are laid down in the 1973 Act. Section 10 states:

"It shall be the duty of each water authority to take all such actions as the authority may from time to time consider necessary or expedient ... for the purposes of conserving, redistributing or otherwise augmenting water resources in their area or of securing the proper use of water resources in their area, or of transferring any such resources to the area of another water authority".

In Scotland, overall responsibility for promoting the conservation of water resources and provision of adequate water supplies lies with the Secretary of State. In Northern Ireland, the Water Service is the responsible authority and is an integral Division of the Department of the Environment for Northern Ireland.

Powers to control the use of ground waters are conferred on water authorities in England and Wales by the Water Resources Act 1963. Ground waters are defined as water contained in any underground strata in that area or water contained in any well, borehole or similar work or any excavation into underground strata where the level of water in the excavation depends wholly or mainly on water entering it from these strata. As with surface waters, rights to abstract ground water for public supplies in Scotland are granted by an order of the Secretary of State under the Water (Scotland) Act 1980, either approving an agreement with the owner or authorizing compulsory acquisition.

Part IV of the 1963 Act provides powers for the water authority to control abstraction of water from any source, including ground water, by means of licences which may be subject to provisions, inter alia, regarding the quantity of water which may be abstracted. Certain exceptions apply to the use of small quantities of water, use by the occupiers of land at the point of abstraction and use as a water supply for domestic purposes. The Act includes provision for disputes involving the issue of licences to be referred to the Minister. There is no licensing arrangement for the control of use of ground water in Scotland. Under the common law, underground water is the property of the owner of the land, except in the case of underground streams passing beneath various properties where individual "riparian" owners have certain rights of abstraction

Water authorities have powers to protect ground waters from pollution under the Control of Pollution Act 1974. Under Section 31, it is an offence to pollute "specified" underground waters which are waters that a water authority declares are being used or may be expected to be used for any purposes. However the Act provides for certain discharges to be authorized subject to a disposal licence or consent issued by the water authority or the Minister, or if the discharge is in accordance with Codes of Good Agricultural Practice issued by the Agricultural Departments.